SpringerBriefs in Food, Health, and Nutrition

Springer Briefs in Food, Health, and Nutritic ng
edge research and practical applications acrc he
field of food science.

Editor-in-Chief

Richard W. Hartel, *University of Wisconsin—Madison, USA*

Associate Editors

J. Peter Clark, *Consultant to the Process Industries, USA*
David Rodriguez-Lazaro, *ITACyL, Spain*
David Topping, *CSIRO, Australia*

For further volumes:
http://www.springer.com/series/10203

Maria Lidia Herrera

Analytical Techniques for Studying the Physical Properties of Lipid Emulsions

Maria Lidia Herrera
University of Buenos Aires
Faculty of Exact and Natural Sciences
Buenos Aires
Argentina

ISBN 978-1-4614-3255-5 e-ISBN 978-1-4614-3256-2
DOI 10.1007/978-1-4614-3256-2
Springer New York Dordrecht Heidelberg London

Library of Congress Control Number: 2012932599

© Maria Lidia Herrera 2012
All rights reserved. This work may not be translated or copied in whole or in part without the written permission of the publisher (Springer Science+Business Media, LLC, 233 Spring Street, New York, NY 10013, USA), except for brief excerpts in connection with reviews or scholarly analysis. Use in connection with any form of information storage and retrieval, electronic adaptation, computer software, or by similar or dissimilar methodology now known or hereafter developed is forbidden.
The use in this publication of trade names, trademarks, service marks, and similar terms, even if they are not identified as such, is not to be taken as an expression of opinion as to whether or not they are subject to proprietary rights.

Printed on acid-free paper

Springer is part of Springer Science+Business Media (www.springer.com)

Contents

Chapter 1
Introduction

Water and lipids are critical for sustaining life and health, but their poor miscibility has posed a challenge for both nature and man. Emulsions are colloidal dispersions that consist of two immiscible liquids, with one of the liquids being dispersed in the other one. Oil dispersions in the form of small spherical droplets are stabilized in the aqueous phase by proteins or surfactants giving an oil-in-water (O/W) emulsion. The surface-active ingredient is adsorbed at the interface between oil and the aqueous phase to lower surface tension and prevent oil droplets from coming close enough together to aggregate.

Emulsions are materials with a widespread range of applications, the most important ones including cosmetics, foods, detergency, adhesives, coatings, paints, surface treatment, road surfacing, and pharmaceutics (Thivilliers et al. 2008). Among edible materials, a considerable number of natural and processed foods consist either partly or wholly as emulsions, or have been in an emulsified state sometime during their production, including milk, cream, fruit beverages, infant formula, soups, cake batters, salad dressings, mayonnaise, cream liqueurs, sauces, desserts, salad cream, ice cream, coffee whitener, spreads, butter, and margarine (McClements 2005a).

The understanding and manipulation of bulk properties of emulsion systems are of the utmost importance in the food industry. The physicochemical and sensory properties of a particular food emulsion depend on the type and concentration of ingredients that it contains, as well as the method used to create it. In addition, shelf life, mouthfeel, and flow properties, to name a few, are to a great extent determined by interactions present among the system's constituents (Corredig and Alexander 2008).

Nanoemulsions can be distinguished from the conventional emulsions that are currently more commonly used in the food industry in terms of their droplet size. Conventional emulsions can be defined as having droplet radii greater than 100 nm, whereas nanoemulsions have radii less than 100 nm (Mason et al. 2006). Conventional emulsions tend to be cloudy or opaque in appearance because the dimensions of the lipid droplets are on the same order as the wavelength of light ($d \approx \lambda$), so that light scattering is relatively strong. On the other hand, nanoemulsions tend to be transparent or slightly turbid in appearance because the dimensions of the lipid droplets are

M.L. Herrera, *Analytical Techniques for Studying the Physical Properties of Lipid Emulsions*, SpringerBriefs in Food, Health, and Nutrition 3,
DOI 10.1007/978-1-4614-3256-2_1, © Maria Lidia Herrera 2012

usually much smaller than the wavelength of light ($d \ll \lambda$), so that light scattering is relatively weak. Conventional emulsions are often prone to gravitational separation and droplet aggregation because of the relatively large size of the droplets. On the other hand, nanoemulsions are usually highly stable to gravitational separation because the relatively small droplet size means that Brownian motion effects dominate gravitational forces. In addition, nanoemulsions tend to have better stability against droplet aggregation than conventional emulsions because the strength of the net attractive forces acting between droplets usually decreases with decreasing droplet size, whereas the strength of the repulsive steric forces is less dependent on size. The rheological properties of nanoemulsions follow similar trends to conventional emulsions; that is, the viscosity increases with increasing droplet concentration and with droplet aggregation. Nevertheless, the viscosity of a nanoemulsion may be appreciably greater than that of a conventional emulsion at the same lipid concentration if it contains a thick or electrically charged interfacial layer that increases droplet–droplet repulsion (McClements and Li 2010).

One of the main properties of a food emulsion is its stability, that is, the ability to resist changes in its properties over time. The length of time that an emulsion must remain stable depends on the nature of the food product. Some food emulsions are formed as intermediate steps during a manufacturing process and therefore only need to remain stable for a few seconds, minutes, or hours (e.g., cake batter, ice cream mix, and margarine premix), whereas others must remain stable for days, months, or even years prior to consumption (e.g., mayonnaise, salad dressings, and cream liqueurs). On the other hand, the production of some foods involves a controlled destabilization of an emulsion during the manufacturing process, for example, margarine, butter, whipped cream, and ice cream (McClements 2005b).

Colloids and dispersions are inherently unstable products, but they can be considered stable if their destabilization velocity is sufficiently low compared with the expected lifespan. Special precautions need to be taken to overcome emulsions' natural tendency to demix or break. A major part of the physical stability of food emulsions is determined by the continuous phase, but also the distribution of droplet sizes plays an important role. Among the major objectives of emulsion scientists working in the food industry is to establish the specific factors that determine the stability of each particular type of food emulsion as well as to elucidate general principles that can be used to predict the behavior of food products or processes. Among physical destabilization mechanisms, the most common phenomena affecting the homogeneity of conventional emulsions are particle migration (creaming, sedimentation) and particle size variation or aggregation (coalescence, flocculation).

In general, the droplets in an emulsion have a different density than that of the liquid that surrounds them, and so a net gravitational force acts on them. The particles float upwards or sink, depending on how large they are and how much less dense or denser they may be than the continuous phase. Creaming is the migration upward of the dispersed phase of an emulsion, while sedimentation is the downward movement of droplets. Gravitational separation is usually regarded as having an adverse effect on the quality of food emulsions. A consumer expects to see a product that appears homogeneous, and therefore the separation of an emulsion into an optically opaque droplet-rich layer and a less opaque droplet-depleted layer is

Fig. 1.1 Confocal laser scattering microscopy (CLSM) images of emulsions with 10 wt.% sunflower oil (SFO) as fat phase and 0.5 wt.% sodium caseinate (NaCas) kept at 22.5°C for 24 h. The *green* color corresponds to the fat phase

undesirable (McClements 2005b). Figure 1.1 shows an example of a confocal laser scanning microscopy image of a sunflower/sodium caseinate emulsion that destabilized by creaming. Emulsion was stabilized by 0.5 wt.% sodium caseinate. Sunflower oil content was 10 wt.%. The image was taken 24 h after preparation. The emulsion was still stable at that time. In an emulsion that mainly destabilized by creaming, the fat phase appeared as individual droplets evenly distributed immediately after preparation; then droplets remained as individual particles during destabilization. Flocculation is a process in which individual particles of a suspension form aggregates, while coalescence is caused by rupture of the film between two emulsion drops or two foam bubbles. The driving force for coalescence is the decrease in free energy resulting when the total surface area is decreased, as occurs after film rupture. The former process is reversible, while the latter leads to bigger particles. Figure 1.2 provides a confocal laser scanning image of a sunflower/sodium caseinate emulsion that destabilized by flocculation. In this case, the sodium caseinate concentration was 5 wt.%. As may be noticed, fat droplets were aggregated, forming flocs. In emulsion systems, the $d_{4,3}$ parameter, the volume-weighted mean diameter of initial emulsions, obtained from droplet size distribution expressed as differential volume, is more sensitive to fat droplet aggregation (coalescence and/or flocculation) than the Sauter mean diameter ($d_{3,2}$) (Relkin and Sourdet 2005). When this sample was analyzed for particle size distribution, immediately after preparation and after a week at 22.5°C, there were no changes in $d_{4,3}$, meaning that the emulsion in the example destabilized by flocculation.

Emulsions may also destabilized by other mechanisms such as partial coalescence, phase inversion, and Ostwald ripening. Partial coalescence is usually described as the destabilization process that occurs when fat crystals present within the thin film separating two droplets pierce the film and bridge the surfaces, causing the

Fig. 1.2 Confocal laser scattering microscopy (CLSM) images of emulsions with 10 wt.% sunflower oil (SFO) as fat phase and 5 wt.% sodium caseinate (NaCas) kept at 22.5°C for 24 h. The *green* color corresponds to the fat phase

droplets to coalesce. If the crystallized fraction within the droplets is sufficient, the intrinsic rigidity inhibits relaxation to the spherical shape driven by surface tension after each coalescence event. Large clusters appear and grow by the accretion of any other primary droplet or cluster until a rigid network made of partially coalesced droplets is formed (Davies et al. 2000; van Aken 2001; Vanapalli and Coupland 2001; Vanapalli et al. 2002; Giermanska-Kahn et al. 2005; Thivilliers et al. 2006; Golemanov et al. 2006; Giermanska et al. 2007; Thivilliers-Arvis et al. 2010).

Emulsion inversion refers to the swap of the dispersed phase and the continuous phase of the emulsion. The phenomenon occurs under certain conditions, and the process is often used as a route to make emulsions (Liu and Friberg 2010). Inversion is a central process in emulsion technology, both as an integral part of the manufacturing technology and as an inevitable component in a large number of emulsion applications, especially those involving evaporation, since the evaporation path, by necessity, leads to inversion when the evaporation takes place predominantly from the continuous phase (Corkery et al. 2010).

Ostwald ripening involves the transport by diffusion of molecules of the disperse phase from small to large particles, due to the differences in Laplace pressure. Laplace stated that the pressure at the concave side of a curved interface is higher than the pressure at the convex side by an amount that is proportional to the interfacial tension times the curvature; for a sphere, the curvature is given by the reciprocal of its radius. The result is that the solubility of the dispersed phase in the continuous phase is increased to a larger extent for a small particle than for a large one; hence, small particles shrink (and will eventually disappear), and large particles grow larger. The rate at which the transport will occur is about proportional to the solubility of the oil in the continuous phase. Given that the solubility of nearly all natural

TAG in most aqueous systems is negligible, the Ostwald ripening is not a common destabilization mechanism in conventional food oil-in-water emulsions (Walstra 2003; Giermanska et al. 2007; Fredrick et al. 2010). Nanoemulsions, however, may be unstable to Ostwald ripening and have to be specifically designed to prevent this phenomenon from occurring, for example, by adding a highly water-insoluble component (Sonneville-Aubrun et al. 2004; Wooster et al. 2008; Li et al. 2009).

As food products are very complex systems, food scientists often rely on the use of analytical techniques to experimentally monitor changes in emulsion properties over time. By using a combination of theoretical understanding and experimental measurements, food manufacturers are able to predict the influence of different ingredients, processing operations, and storage conditions on the stability and properties of food emulsions (McClements 2005b). In this book, the physical and chemical properties of emulsions are described with a special focus on food emulsions' stability. The experimental techniques for monitoring emulsion stability, the major factors that influence the different destabilization mechanisms, as well as methods of controlling them are discussed.

References

Corkery RW, Blute IA, Friberg SE, Guo R (2010) Emulsion inversion in the PIT range: quantitative phase variations in a two-phase emulsion. J Chem Eng Data 55:4471–4475

Corredig M, Alexander M (2008) Food emulsions studied by DWS: recent advances. Trends Food Sci Technol 19:67–75

Davies E, Dickinson E, Bee R (2000) Shear stability of sodium caseinate emulsions containing monoglyceride and triglyceride crystals. Food Hydrocolloids 14:145–153

Fredrick E, Walstra P, Dewettinck K (2010) Factors governing partial coalescence in oil-in-water emulsions. Adv Colloid Interface Sci 153:30–42

Giermanska J, Thivilliers F, Backov R, Schmitt V, Drelon N, Leal-Calderon F (2007) Gelling of oil-in-water emulsions comprising crystallized droplets. Langmuir 23:4792–4799

Giermanska-Kahn J, Laine V, Arditty S, Schmitt V, Leal-Calderon F (2005) Particle-stabilized emulsions comprised of solid droplets. Langmuir 21:4316–4323

Golemanov K, Tcholakova S, Denkov ND, Gurkov T (2006) Selection of surfactants for stable paraffin-in-water dispersions, undergoing solid-liquid transition of the dispersed particles. Langmuir 22:3560–3569

Li Y, Le Maux S, Xiao H, McClements DJ (2009) Emulsion-based delivery systems for tributyrin, a potential colon cancer preventative agent. J Agric Food Chem 57:9243–9249

Liu Y, Friberg SE (2010) Perspectives of phase changes and reversibility on a case of emulsion inversion. Langmuir 26:15786–15793

Mason TG, Wilking JN, Meleson K, Chang CB, Graves SM (2006) Nanoemulsions: formation, structure, and physical properties. J Phys Condens Matter 18:R635–R666

McClements DJ (2005a) Context and background. In: Food emulsions, principles, practices, and techniques, 2nd edn. CRC Press, New York, pp 1–26

McClements DJ (2005b) Emulsion stability. In: Food emulsions, principles, practices, and techniques, 2nd edn. CRC Press, New York, pp 269–339

McClements DJ, Li Y (2010) Structured emulsion-based delivery systems: controlling the digestion and release of lipophilic food components. Adv Colloid Interface Sci 159:213–228

Relkin P, Sourdet S (2005) Factors affecting fat droplet aggregation in whipped frozen protein-stabilized emulsions. Food Hydrocolloids 19:503–511

Sonneville-Aubrun O, Simonnet JT, L'Alloret F (2004) Nanoemulsions: a new vehicle for skincare products. Adv Colloid Interface Sci 108:145–149
Thivilliers F, Drelon N, Schmitt V, Leal-Calderon F (2006) Bicontinuous emulsion gels induced by partial coalescence: kinetics and mechanism. Europhys Lett 76:332–338
Thivilliers F, Laurichesse E, Saadaoui H, Leal-Calderon F, Schmitt V (2008) Thermally induced gelling of oil-in-water emulsions comprising partially crystallized droplets: the impact of interfacial crystals. Langmuir 24:13364–13375
Thivilliers-Arvis F, Laurichesse E, Schmitt V, Leal-Calderon F (2010) Shear-induced instabilities in oil-in-water emulsions comprising partially crystallized droplets. Langmuir 26:16782–16790
van Aken GA (2001) Aeration of emulsions by whipping. Colloids Surf A Physicochem Eng Asp 190:333–354
Vanapalli SA, Coupland JN (2001) Emulsions under shear. The formation and properties of partially coalesced lipid structures. Food Hydrocolloids 15:507–512
Vanapalli SA, Palanuwech J, Coupland JN (2002) Stability of emulsions to dispersed phase crystallization: effect of oil type, dispersed phase volume fraction, and cooling rate. Colloids Surf A Physicochem Eng Asp 204:227–237
Walstra P (2003) Crystallization of lipids. In: Physical chemistry of foods. Marcel Dekker, New York, pp 476–547
Wooster TJ, Golding M, Sanguansri P (2008) Impact of oil type on nanoemulsion formation and Ostwald ripening stability. Langmuir 24:12758–12765

Chapter 2
Nano and Micro Food Emulsions

2.1 Methods of Formation

Emulsions with microdroplets, sometimes called *conventional emulsions*, and nanodispersions, or thermodynamically stable emulsions (surprisingly called *micro-emulsions*), can be easily manufactured on an industrial scale up. Due to their satisfactory stability over a certain storage time and high bioavailability, they have attained particular interest as delivery systems for bioactive substances, such as carotenoids, phytostetol, polyunsaturated fatty acids, g-oryzanol, lipophilic vitamins, and numerous other compounds. Garti and co-workers (Amar et al. 2003; Spernath et al. 2002), for example, prepared food-grade conventional emulsions containing carotenoids with considerable success. Recently, studies have shown the successful approach of using nanoemulsions to improve stability in food applications. Tan (2005) and Yuan et al. (2008) prepared β-carotene nanodispersions using high-pressure homogenization and studied their physicochemical properties. Other applications include the encapsulation of limonene, lutein, omega-3 fatty acids, astaxantin, and lycopene (Chen et al. 2006), the encapsulation of α-tocopherol to reduce lipid oxidation in fish oil (Weiss et al. 2006), and the use of nanoemulsions to incorporate essential oils, oleoresins, and oil-based natural flavors into food products such as carbonated beverages and salad dressings (Ochomogo and Monsalve-Gonzalez 2009).

2.1.1 Nanoemulsions

In the last two decades, nanotechnology has rapidly emerged as one of the most promising and attractive research fields. The technology offers the potential to significantly improve the solubility and bioavailability of many functional ingredients. The high hydrophobicity of some bioactive substances makes them insoluble in aqueous systems, and they therefore have a poor intake in the body. To improve

M.L. Herrera, *Analytical Techniques for Studying the Physical Properties of Lipid Emulsions*, SpringerBriefs in Food, Health, and Nutrition 3,
DOI 10.1007/978-1-4614-3256-2_2, © Maria Lidia Herrera 2012

carotenoids' dispersibility in water, for example (Horn and Rieger 2001), and also to increase their bioavailability during gastrointestinal passage (Deming and Erdman 1999), carotenoid crystals must be formulated, that is, to incorporate them in the fine particles of oil-in-water (O/W) emulsions.

Nanoemulsions are nonequilibrium systems and cannot be formed spontaneously. They can be produced using two different approaches: high-energy and low-energy methods. High-energy methods use intense mechanical forces to break up macroscopic phases or droplets into smaller droplets and typically involve the use of mechanical devices known as homogenizers, which may use high-shear mixing, high-pressure homogenization, or ultrasonification. In contrast, low-energy methods rely on the spontaneous formation of emulsions under specific system compositions or environmental conditions as a result of changes in interfacial properties (McClements and Li 2010). Spontaneous emulsification is a less expensive and energy-efficient alternative that takes advantage of the chemical energy stored in the system (Bilbao Sáinz et al. 2010). High-energy methods are effective in reducing droplet sizes but may not be suitable for some unstable molecules, such as proteins or peptides.

One of the most used low-energy methods is the phase inversion temperature (PIT) method, which is based on the changes in solubility of polyoxyethylene-type nonionic surfactants with temperature. The surfactant is hydrophilic at low temperatures but becomes lipophilic with increasing temperature due to dehydration of the polyoxyethylene chains. At low temperatures, the surfactant monolayer has a large positive spontaneous curvature forming oil-swollen micellar solution phases (or O/W microemulsions), which may coexist with an excess oil phase. At high temperatures, the spontaneous curvature becomes negative and water-swollen reverse micelles (or W/O microemulsions) coexist with an excess water phase. At a critical temperature—the hydrophilic–lipophilic balance (HLB) temperature—the spontaneous curvature is zero and a bicontinuous microemulsion phase containing comparable amounts of water and oil phases coexists with both excess water and oil phases. The PIT emulsification method takes advantage of the extremely low interfacial tensions at the HLB temperature to promote emulsification. However, the coalescence rate is extremely fast and the emulsions are very unstable even though emulsification is spontaneous at the HLB temperature. By rapidly cooling or heating the emulsions prepared at the HLB temperature, kinetically stable emulsions (O/W or W/O, respectively) can be produced with a very small droplet size and narrow size distribution. If the cooling or heating process is not fast, coalescence predominates and polydispersed coarse emulsions are formed (Ee et al. 2008).

Other low-energy methods that can be used to prepare nanoemulsions are the PIC (phase inversion composition) method, in which the temperature is maintained constant and the composition is changed (the solvent quality is changed by mixing two partially miscible phases together). By using the PIC method, different nanomaterials such as colloidosomes (Dinsmore et al. 2002), cubosomes (Spicer 2004), and microfluidic channels (Xu et al. 2005) have been prepared. Liu et al. (2006) used polyoxyethylene (PEO) nonionic surfactants for the preparation of paraffin oil-in-water nanoemulsions also by using the PIC method. They reported the preparation of stable nanoemulsions with diameters ranging from 100 to 200 nm. Surfactants used in this system were a combination of Span 80, a sorbitan monooleate with a low HLB (4.3), and Tween 80, an ethoxilated sorbitan monooelate ester with a high HLB. As these two surfactants possess the same backbone, they can mix easily, leading to a controlled change in the final HLB.

Low-energy emulsification also includes the catastrophic inversion method (CPI). In a typical phase inversion process, emulsification starts with a given emulsion morphology that inverts to an opposite emulsion by variations in emulsion properties. For example, an oil-in-water system (O/W) inverts into a water-in-oil system (W/O) and vice versa. Catastrophic inversion is induced by increasing the volume fraction of the dispersed phase. In CPI emulsification, the system usually begins with abnormal emulsions, that is, emulsions in which the surfactant has a high affinity to the dispersed phase. Abnormal emulsions are usually unstable and can only be maintained under vigorous mixing for a short period of time. The ultimate fate of an unstable emulsion is to invert to the opposite state. A CPI is triggered by increasing the rate of droplet coalescence so that the balance between the rate of coalescence and rupture cannot be maintained. This may be induced by changing the variables that increase the rate of droplet coalescence, such as the continuous addition of dispersed-phase volume, the most common variable used, or by adding a surfactant or a salt, or by altering the emulsification temperature or any parameter that can significantly enhance droplets' coalescence. Droplets formed via CPI are usually above micrometer. Submicrometer droplets may only form if CPI of abnormal to normal emulsion occurs in the vicinity of the locus of ultralow interfacial tension (Sajjadi et al. 2004; Jahanzad et al. 2010). Peng et al. (2010) adopted a low-energy method combining the PIT and CPI methods to prepare the water-in-oil nanoemulsion. The aim of their work was to gain a better understanding of the relationship among the ratio of surfactants, the water/oil ratio, and long-term stability. They found that the addition of a second surfactant to the formulation could provide more stable nanoemulsions with the minimum size than only one surfactant. This result was in agreement with other authors' findings, who reported that to form nanoemulsions, surfactant mixtures generally perform better than pure surfactants for various applications (Rees et al. 1999; Uskokovi and Drofenik 2005; Pey et al. 2006). Peng et al. (2010) also reported the optimum composition for the systems they studied.

Another commonly used inversion method is transitional phase inversion (TPI) (Jahanzad et al. 2010). Before TPI can occur, the surfactants in both phases must diffuse toward the interface, adsorb at the interface, and conform into a mixed surfactant layer at the optimum conditions. The rate of diffusion of the surfactants depends on many parameters, including their size, the viscosity of the phase, and the intensity of mixing, etc. For oil-in-water emulsions containing a pair of water-soluble and oil-soluble surfactants, it was found that the addition of the water phase containing the water-soluble surfactant to the oil phase containing the oil-soluble surfactant may produce very fine emulsions if it is associated with interfacial tension lowering in the course of addition. The rate of addition of the second phase is of great importance in achieving an emulsion with the desired properties. This is because the dynamic of phase inversion emulsification is very fast, and the emulsion properties change quickly with further addition, contrary to some conventional emulsification methods. Therefore, it is important to find and maintain a semi-equilibrated state in the course of emulsification during which sufficient surfactant diffusion/adsorption can occur, and thus droplet rupturing is enhanced. TPI occurs when the curvature of the oil–water interface gradually changes from positive to negative, passing through a zero curvature at the inversion point. This is associated with a shift in the surfactant nature from water-soluble to oil-soluble, or vice versa. At the inversion point, the surfactant has a similar affinity toward both phases. As a result, the interfacial tension passes through an ultralow value. This results in the formation of emulsions with a very small drop size, sometimes called *miniemulsions* and *nanoemulsions* (Jahanzad et al. 2010).

In a recent work, Ribeiro et al. (2008) produced β-carotene-loaded nanodispersions containing poly(D,L-lactic acid) (PLA) and poly(D,L-lactic-coglycolic acid) (PLGA) by the solvent displacement method. Nanoparticles containing β-carotene were produced by interfacial deposition of the biodegradable polymer, due to the displacement of acetone from the dispersed phase. Gelatin or Tween 20 was used as stabilizing hydrocolloids in the continuous phase. β-carotene was entrapped in the polymeric matrix in the absence of any oily core material. In this kind of formulation, polymers assumed the function of protective colloids and possibly also chemical stabilization of the nanodispersed phase.

2.1.2 Conventional Emulsions

Conventional emulsions, that is, emulsions with microdroplets, are thermodynamically unstable liquid–liquid dispersions. Emulsions with particle sizes higher than 100 nm are usually prepared by high-energy methods. As in the case of nanoemulsions, the O/W systems consist of lipid droplets dispersed in an aqueous medium, with each lipid droplet being surrounded by a thin emulsifier layer. The initial droplet concentration and size distribution of this type of delivery system can be controlled, as can the nature of the emulsifier used to stabilize the droplets. A careful selection of emulsifier type enables one to control interfacial properties such as charge, thickness (dimensions), rheology, and response to environmental stresses (such as pH, ionic strength, temperature, and enzyme activity) (McClements and Li 2010). To form conventional emulsions, usually a preemulsion with coarse particles, obtained by a mechanical device such as an Ultra-Turrax, is prepared. This system is further homogenized to obtain a microdroplet emulsion. In general, a variety of homogenizers are available to prepare emulsions, including high-shear mixers, high-pressure homogenizers, colloid mills, ultrasonic homogenizers, and membrane homogenizers. The choice of a particular kind of homogenizer and of the operating conditions used depends on the characteristics of the materials being homogenized (e.g., viscosity, interfacial tension, and shear sensitivity) and of the required final properties of the emulsion (e.g., droplet concentration, droplet size, and viscosity) (McClements

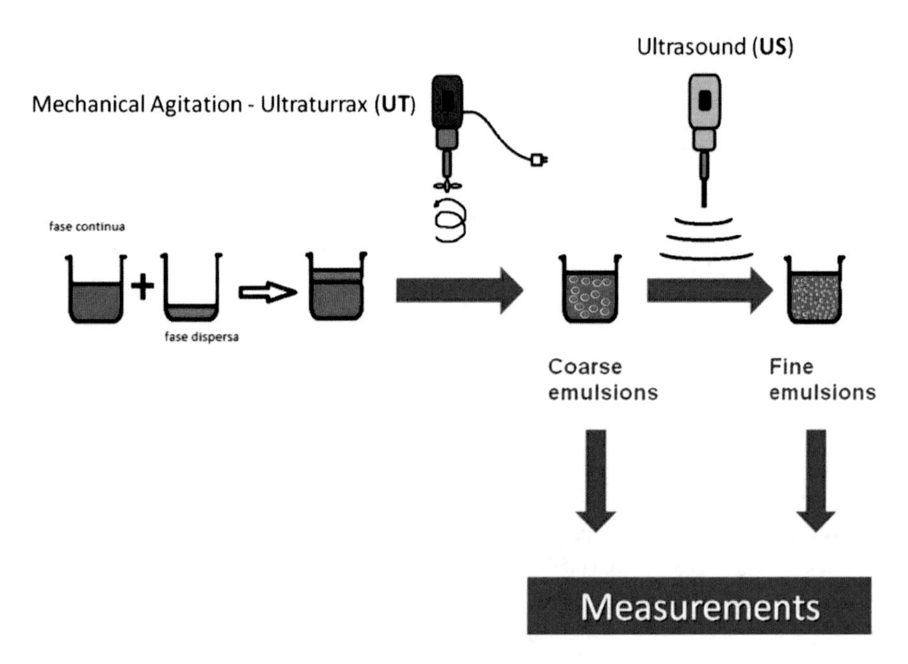

Fig. 2.1 Scheme of conventional emulsions preparation procedure

and Li 2010). Figure 2.1 illustrates a preparation procedure. This is a common approach widely used in the literature. As an example, the typical conditions are as follows: The fat and aqueous phases may be mixed using an Ultra-Turrax high-speed blender operated at 20,000 rpm for 1 min. The resulting preemulsions may be further homogenized for 20 min using an ultrasonic liquid processing. It is advisable that the temperature of the sample cell is controlled by means of a water bath usually set at a temperature that prevents protein denaturalization and that does not increase the system viscosity very much, such as 15°C. By doing this, the sample temperature never rises higher than 40°C during ultrasound treatment. Then emulsions are typically cooled quiescently to ambient temperature (22.5°C) to perform physicochemical analysis. Particle sizes obtained with this protocol may vary from 0.2 to 1 μm depending on the system under study (Álvarez Cerimedo et al. 2010).

2.2 Physical Chemical Properties

2.2.1 Nanoemulsions

The physical properties of nanoemulsions can be characterized by the combination of a wide variety of techniques. For example, the macroscopic properties, such as viscosity/viscoelasticity, conductivity, and interfacial tension, can be characterized by rheometer, conductivity meter, and pendant drop tensiometer, respectively (Boonme et al. 2006). The size and shape of the emulsion droplets were routinely characterized by static and dynamic light scattering techniques (McClements 2005). The major drawback of light scattering techniques is that dilution of emulsion samples is usually necessary to reduce multiple scattering and interdroplet interactions. The dilution process may modify the structure and composition of the pseudoternary phases of the nanoemulsions; therefore, the obtained results do not accurately describe the actual system under study (Huang et al. 2010).

2.2.2 *Conventional Emulsions*

Conventional emulsions may also be characterized by the techniques mentioned above. However, there are other alternatives that allow emulsions to be described without the drawback of dilution. For example, particle size may be measured by low-resolution nuclear magnetic resonance (NMR). As will be explained in Chap. 3, the principle of this technique is completely different than that of light scattering techniques, and in some cases, a more accurate description of the system may therefore be made.

2.3 Structuring Food Emulsions

The structure of the different pseudoternary phases can be investigated by small-angle X-ray scattering (SAXS), small-angle neutron scattering (SANS), and microscopy-like cryotransmission electron microscopy (TEM) (Spicer et al. 2001; Borne et al. 2002; Boonme et al. 2006). These techniques may also be used to describe conventional emulsions' structure. By using SAXS measurements, Álvarez Cerimedo et al. (2010) proved that the role of trehalose in caseinate/fish oil emulsions went beyond the ability to form viscous solutions. For those systems, values of q (the reciprocal lattice spacing, with $q=2\pi/d=4\pi\ \sin(\theta)/\lambda$, where d is the interplanar spacing and 2θ is the Bragg angle) were significantly modified when the aqueous phase contained trehalose compared to emulsions without sugar in the aqueous phase. Values for emulsions with 10 wt.% fish oil and 5 wt.% sodium caseinate were 241, 0.248, and 0.252 nm^{-1} for emulsions with 0, 20, and 30 wt.% trehalose, respectively. Some other aqueous phase components such as hydrocolloids proved to stabilize emulsions because they increase viscosity. The slightly increased q values with the trehalose addition suggested that trehalose had an effect beyond viscosity changes since modification of q values means that the aggregation state of the protein changed with the aqueous phase formulation. These results were in agreement with the small particle size as measured by dynamic light scattering found when trehalose was added to aqueous phase. These results were corroborated by confocal laser scanning microscopy (CLSM). CLSM is a more common technique to describe emulsion structure. It is widely used in food applications and provides a good description of the spatial distribution of different phases.

References

Álvarez Cerimedo MS, Huck Iriart C, Candal RJ, Herrera ML (2010) Stability of emulsions formulated with high concentrations of sodium caseinate and trehalose. Food Res Int 43:1482–1493

Amar I, Aserin A, Garti N (2003) Solubilization patterns of lutein and lutein esters in food grade nonionic microemulsions. J Agric Food Chem 51:4775–4781

Bilbao Sáinz C, Avena Bustillos RJ, Wood DF, Williams TG, Mchugh TH (2010) Nanoemulsions prepared by a low-energy emulsification method applied to edible films. J Agric Food Chem 58:11932–11938

Boonme P, Krauel K, Graf A, Rades T, Junyaprasert VB (2006) Characterization of microemulsion structures in the pseudoternary phase diagram of isopropyl palmitate/water/Brij 97:1-butanol. AAPS Pharm Sci Technol 7:1–6

Borne J, Nylander T, Khan A (2002) Effect of lipase on monoolein-based cubic phase dispersion (cubosomes) and vesicles. J Phys Chem B 106:10492–10500

Chen H, Weiss J, Shahidi F (2006) Nanotechnology in nutraceuticals and functional foods. Food Technol 60:30–36

Deming DM, Erdman JW Jr (1999) Mammalian carotenoid absorption and metabolism. Pure Appl Chem 71:2213–2223

Dinsmore AD, Hsu MF, Nikolaides MG, Manuel M, Bausch AR, Weitz DA (2002) Colloidosomes: selectively permeable capsules composed of colloidal particles. Science 298:1006–1009

Ee SL, Duan X, Liew J, Nguyen QD (2008) Droplet size and stability of nano-emulsions produced by the temperature phase inversion method. Chem Eng J 140:626–631

Horn D, Rieger J (2001) Organic nanoparticles in the aqueous phase—theory, experiment, and use. Angewandte Chemie 113:4460–4492

Huang QR, Yu HL, Ru QM (2010) Bioavailability and delivery of nutraceuticals using nanotechnology. J Food Sci 75:R50–R57

Jahanzad F, Josephides D, Mansourian A, Sajjadi S (2010) Dynamics of transitional phase inversion emulsification: effect of addition time on the type of inversion and drop size. Ind Eng Chem Res 49:7631–7637

Liu W, Sun D, Li C, Liu Q, Xu JJ (2006) Formation and stability of paraffin oil-in-water nanoemulsion prepared by the emulsion inversion point method. J Colloid Interface Sci 303:557–563

McClements DJ (2005) Emulsion stability. In: Food emulsions: principles, practices, and techniques, 2nd edn. CRC Press, New York, pp 269–339

McClements DJ, Li Y (2010) Structured emulsion-based delivery systems: controlling the digestion and release of lipophilic food components. Adv Colloid Interface Sci 159:213–228

Ochomogo M, Monsalve-Gonzalez A (2009) Natural flavor enhancement compositions for food emulsions. U.S. Patent 2009/0196972 A1. Clorox Co., Oakland

Peng LC, Liu CH, Kwan CC, Huang KF (2010) Optimization of water-in-oil nanoemulsions by mixed surfactants. Colloids Surf A Physicochem Eng Asp 370:136–142

Pey CM, Maestro A, Solé I, González C, Solans C, Gutiérrez JM (2006) Nano-emulsions preparation by low energy methods in an ionic surfactant system. Colloids Surf A Physicochem Eng Asp 288:144–150

Rees GD, Evans-Gowing R, Hammond SJ, Robinson BH (1999) Formation and morphology of calcium sulfate nanoparticles and nanowires in water-in-oil microemulsions. Langmuir 15:1993–2002

Ribeiro HS, Chu BS, Ichikawa S, Nakajima M (2008) Preparation of nanodispersions containing β-carotene by solvent displacement method. Food Hydrocolloids 22:12–17

Sajjadi S, Jahanzad F, Yianneskis M (2004) Catastrophic phase inversion of abnormal emulsions in the vicinity of the locus of transitional inversion. Colloids Surf A Physicochem Eng Asp 240:149–155

Spernath A, Yaghmur A, Aserin A, Hoffman RE, Garti N (2002) Microemulsions studied by self-diffusion NMR. J Agric Food Chem 50:6917–6922

Spicer PT (2004) Cubosomes: bicontinuous liquid crystalline nanoparticles. In: Nalwa H (ed) Encyclopedia of nanoscience and nanotechnology. Marcel Dekker, New York, pp 881–892

Spicer PT, Hayden KL, Lynch ML, Ofori-Boateng A, Burns JL (2001) Novel process for producing cubic liquid crystalline nanoparticles (cubosomes). Langmuir 17:5748–5756

Tan CP (2005) Effect of polyglycerol esters of fatty acids on physicochemical properties and stability of β-carotene nanodispersions prepared by emulsification/evaporation method. J Sci Food Agric 85:121–126

Uskokovi V, Drofenik M (2005) Synthesis of materials within reverse micelles. Surface Rev Lett 12:239–277

Weiss J, Takhistov P, McClements J (2006) Functional materials in food nanotechnology. J Food Sci 71:R107–R116

Xu QY, Nakajima M, Binks BP (2005) Preparation of particle-stabilized oil-in-water emulsions with the microchannel emulsification method. Colloids Surf A Physicochem Eng Asp 262:94–100

Yuan Y, Gao Y, Zhao J, Mao L (2008) Characterization and stability evaluation of β-carotene nanoemulsions prepared by high pressure homogenization under various emulsifying conditions. Food Res Int 41:61–68

Chapter 3
Methods for Stability Studies

The efficient development and production of high-quality emulsion-based products depend on knowledge of their physicochemical properties and stability. A wide variety of different analytical techniques and methodologies have been developed to characterize the properties of food emulsions. Analytical instruments and experimental methodologies are needed for research and development purposes to elucidate the relationship between droplet characteristics and the bulk physicochemical and sensory properties of food emulsions, such as stability, texture, flavor, and appearance. They are also needed in quality control laboratories and in food production factories to monitor food emulsions and their components before, during, and after production so as to ensure that their properties conform to predefined quality criteria and/or to predict how the final product will behave during storage. This chapter describes the most commonly used methods for stability studies, with a focus on conventional food emulsions. Some of these techniques are also used in nanoemulsions. Several examples of applications are described in detail.

3.1 Visual Observation

Emulsion systems have minimal thermodynamic stability and tend to phase-separate. The primary driving force for phase separation is droplet interfacial free energy. The inclusion of a surface-active substance that concentrates at the oil–water interface imparts a degree of stability to these systems by lowering the interfacial tension. In addition, a reduction in the interfacial tension facilitates emulsion formation and prevents immediate droplet recoalescence during preparation. Visual observation is an old method still used to analyze emulsion stability. Emulsion instability is studied by placing the samples in tubes, which are stored in quiescent conditions. Destabilization is commonly indicated by the separation of water phase at the bottom of the container or by a complete breakdown into two phases with a layer of surfactant in between; that is, there is an opaque layer at the top, a turbid layer in the

M.L. Herrera, *Analytical Techniques for Studying the Physical Properties of Lipid Emulsions*, SpringerBriefs in Food, Health, and Nutrition 3,
DOI 10.1007/978-1-4614-3256-2_3, © Maria Lidia Herrera 2012

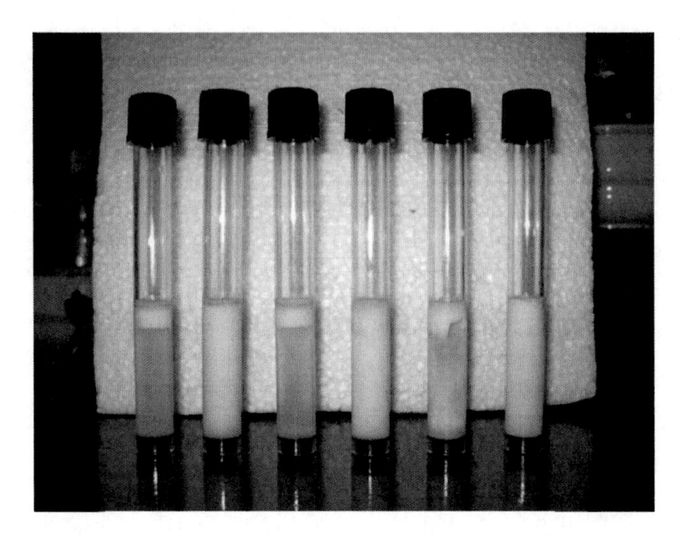

Fig. 3.1 Visual aspect of emulsions formulated with 0.3 wt.% of locust bean gum in the aqueous phase and 10 wt.% of sunflower oil as the fat phase stabilized with different concentrations of sodium caseinate (NaCas) and stored in quiescent conditions for a week at 22.5°C. From *left* to *right*: coarse emulsion stabilized with 0.5 wt.% NaCas; fine emulsion stabilized with 0.5 wt.% NaCas; coarse emulsion stabilized with 2 wt.% NaCas; fine emulsion stabilized with 2 wt.% NaCas; coarse emulsion stabilized with 5 wt.% NaCas; and fine emulsion stabilized with 5 wt.%

middle, and/or a transparent layer at the bottom. The serum layer may be defined as the sum of the turbid and transparent layers (Mun et al. 2008), and creaming kinetics may be followed by measuring the total height of the emulsion (HE) and the height of the serum layer (HS). The extent of creaming may be characterized by a creaming index as follows:

$$\mathrm{CI} = 100 \times (\mathrm{HS} / \mathrm{HE})$$

The creaming index is an indirect indication of the extent of droplet aggregation in an emulsion (Mun et al. 2008). Figure 3.1 shows sample emulsions with different creaming indexes.

Although visual observation provides limited information about food emulsion stability, there are still new studies in the literature that are a contribution to their fields. Some of them are mentioned here as an example.

Emulsifiers, usually small molecule surfactants, are added to the system to slow emulsion-breaking mechanisms such as creaming and coalescence. The use of polymeric surfactants as emulsifiers and stabilizers in emulsions has attracted much attention in recent years. Among the polymeric surfactants, hydrophobically modified water-soluble polymers have been a target of extensive studies because of their potential industrial applications such as waterborne paints, coating fluids, cosmetics, foodstuff, drug delivery systems, oil recovery, and water treatment and also because of their relevance to biological macromolecular systems. This kind of polymer is a water-soluble polymer with a few hydrophobic groups in the hydrophilic macromolecular chain. They can lower the surface and interfacial tension due to their intrinsic amphiphilic properties, though their abilities are much lower than those of the conventional low-molecular-weight surfactants. However, their solutions showed much better thickening ability than low-molecular-weight surfactants due to their large molecular structures and the association of the hydrophobic groups in the backbone. Hence, the emulsification mechanisms differ from those of the low-molecular-weight surfactants. Sun et al. (2007) studied the stability of emulsions formulated with a hydrophobically modified hydroxyethyl cellulose (HMHEC) and paraffin oil. The stability and droplet size distribution were investigated by visual observation, photomicrograph, and a laser scattering particle size distribution analyzer. Then the emulsions were transferred to a glass tube with a diameter of 1.6 cm and length of 15 cm to monitor their stability to creaming and coalescence. The stability of emulsion with time was accessed by monitoring the variation of the emulsion volume. The adsorption of HMHEC at the oil–water interface and the surface of emulsion droplets due to the penetration of the alkyl chains in HMHEC into the oil phase was confirmed by visual observation, the interfacial tension method, and an in situ environmental scanning electron microscope (ESEM). According to the authors, the stability of emulsions prepared using HMHEC is based on both an associative thickening mechanism caused by alkyl chains in HMHEC and the adsorption of HMHEC at the oil–water interface, which can form a solid film, preventing coalescence of the droplets.

Proteins and polysaccharides are the two most important biopolymers used in food emulsions to control their texture, microstructure, and stability. Polysaccharides are mainly added to enhance the viscosity or to stabilize the system by the formation of a gel, while proteins can form networks and have emulsification and foaming properties (Dickinson 2003). Network formation and the phase properties of mixed biopolymers affect the rheology of the system; therefore, a considerable research effort has been devoted to understanding the mechanisms of network formation and phase separation. Khalloufi et al. (2008) studied the behavior of two varieties of flaxseed gums when added to whey protein isolate (WPI)-coated emulsion droplets at neutral pH. Flaxseeds (*Linum usitatissimum*) have received increasing attention because of their recognized health benefits, mainly related to the high level of ω-3 fatty acids in the oil. A gum extract is obtained as the byproduct of the oil extraction, and although quite rich in functional components, it is usually processed as livestock feed. Flaxseed gum consists of a mixture of water-soluble polysaccharides with a composition that varies depending on the extraction conditions. It is composed mainly of L-galactose, D-xylose, L-rhamnose, and D-galacturonic acid. Concentrations ranging from 0% to 0.33% (w/v) of gum were added to the emulsions at pH 7. At low concentrations [gum $\leq$ 0.075% (w/v)], no visual phase separation was observed. However, at high concentrations of gum ($\geq$0.1 w/v), there was visual phase separation over time. Khalloufi et al. (2008) demonstrated that flaxseed gum is a non-interacting polysaccharide at neutral pH; it could then be employed to strengthen the nutritional value of some milk-based drinks, but at limited concentrations. To further investigate the mechanisms of network formation and phase separation in polysaccharide/protein systems, Moschakis et al. (2010) analyzed whey protein-stabilized/

sunflower oil emulsions in the presence of polysaccharide mixtures, using chitosan and gum arabic at different weight ratios, and examined emulsions' behavior in terms of their stability, microstructure, and functional properties. Stability was followed by visual observation. Freshly prepared emulsions were poured into 5-mL glass tubes (height=75 mm, diameter=9 mm) after preparation. Subsequently, the tubes were sealed to prevent evaporation. The tubes were then inverted carefully several times to ensure thorough mixing. The emulsion samples were stored quiescently at ambient temperature and the movement of any creaming boundaries was followed with time. The emulsion's stability, properties, and microstructure were found to be greatly dependent on the precise gum arabic-to-chitosan ratio. Mixing of gum arabic with chitosan led to the formation of coacervates of a size dependent on their ratio. The incorporation of low gum arabic-to-chitosan weight ratios into emulsions caused depletion flocculation and gravity-induced phase separation. Increasing the polysaccharide weight ratio further generated a droplet network with a rather high viscosity (at low-shear stress), which prevented or even inhibited phase separation. At even higher gum arabic-to-chitosan ratios, the emulsion droplets were immobilized into clusters of an insoluble ternary matrix.

The addition of the polymer to a dispersion can promote stability or destabilize the dispersion, depending on the nature of interactions between the polymer and the solvent and between the polymer and the dispersed particles (Hiemenz and Rajagopalan 1997). The previous example and other studies with caseins that were reported in the literature provide experimental support to Hiemenz and Rajagopalan's theory. According to this theory, usually called *the polymer theory*, some of the possible effects of polymer chains on a dispersion may be summarized as follows: In the case of very low polymer concentrations, bridging flocculation may occur as a polymer chain forms bridges by adsorbing on more than one particle. At higher concentrations, distances among particles may be higher and mask van der Waals' attraction between the particles, causing steric stabilization. At moderate to high polymer concentrations, the free polymer chains in the solution cause depletion flocculation. At even higher polymer concentrations, the effect is known as *depletion stabilization*. The polymer-depleted regions between the particles can only be created by demixing the polymer chains and solvent. In good solvents, the demixing process is thermodynamically unfavorable, and under such conditions one can have depletion stabilization (Hiemenz and Rajagopalan 1997).

Today the food industry has a growing interest in the replacement of synthetic emulsifiers by natural ones, such as polysaccharides and proteins. Among proteins, the caseins, a group of unique milk-specific proteins that represent around 80% of the total protein in the milk of cattle and other commercial dairying species, are the most studied systems. Casein is widely used as an ingredient in the food industry; its functional properties include emulsification, water-binding, fat-binding, thickening, and gelation. Depending on the product, the caseinate content can range anywhere from <1% to >50%. Casein is of particular importance as an emulsifier because of its ability to rapidly confer a low interfacial tension during emulsification and because of the strongly amphiphilic characteristics of the major individual caseins. Casein in milk is strongly aggregated into polydispersed protein particles called *casein micelles*. The casein micelles are colloidal particles of self-associated casein held together by nanometer-sized clusters of calcium phosphate and sterically stabilized by an outer layer of the glycoprotein *k*-casein. Early electron microscopy studies showed that the micelles have an uneven, raspberry-like appearance, which was interpreted to mean that the micelles are built up from submicelles. However, in recent studies, the irregularities were considered to be microtubules (Dalgleish et al. 2004). The micelles may be dispersed by adding a calcium chelator or also by urea, SDS, high pH, or ethanol, indicating that hydrogen bonds, hydrophobic interactions, and electrostatic interactions are also involved in micelle integrity. The caseins have been described as rheomorphic proteins, indicating that they adopt molecular structures in solution dictated by the local environment; that is, their structures are flexible enough to "go with the flow." Removing the calcium salts from milk casein and replacing them with sodium salts leads to the production of "sodium caseinate" (NaCas). Commercial NaCas is a variable multicomponent mixture containing four major constituents, α_{s1}, α_{s2}, β, and *k*-casein. NaCas is composed of a soluble mixture of disordered hydrophobic proteins having a strong tendency to associate into small protein particles (casein submicelles, $\approx 2.5\times10^5\,\text{Da}$), which coexist in equilibrium with the free casein molecules ($\approx 2.5\times10^4\,\text{Da}$). In the casein micelle system, the micelle state may be the lowest free-energy state of the system (Fox and Brodkorb 2008). Of particular interest will be micelle structure and the mechanisms that operate in determining micelle size since they are closely related to emulsion stability. Selecting sodium caseinate as a stabilizer, Matsumiya et al. (2011) studied how the structural relationship between oil molecules and fatty acid residues of emulsifiers affects the stability of its emulsions. Sodium caseinate–based emulsions were formulated with seven different oils (corn oil, soybean oil, and five hydrocarbons, C10H22, C12H26, C14H30, C16H34, C18H38) and diglycerol monoesters of different saturated fatty acids or one mono-unsaturated fatty acid (lauric, myristic, palmitic, stearic, or oleic acid). Emulsion destabilization, such as creaming, coalescence, or phase separation, caused by diglycerol esters of fatty acids was visually observed. These destabilization phenomena were recorded by taking pictures from 0 min (just after adding emulsifiers) to 60 min. According to these authors, the molecular structural similarity between dispersed oil molecules and emulsifiers, i.e., the similarity of carbon chain length between hydrocarbon oil molecules and fatty acid residues of emulsifiers, could affect the emulsion stability.

Collagen is a protein widely used in food industries to improve the elasticity, consistency, and stability of foods, but this use has only been carried out in an empirical way. A number of ingredients are obtained from collagen, such as gelatin and collagen fiber. Collagen fiber is a new ingredient obtained from collagen in its crude form, while gelatin is produced from the hydrolysis of collagen under drastic conditions. Gelatin is widely used as a gelling agent, but it shows weaker emulsifying properties than other surface-active biopolymers such as globular proteins. In this context, de Castro Santana et al. (2012) investigated the effects of heat treatment on the emulsifying properties of collagen fiber. First, the heat-treated collagen fiber was characterized, and then the effect of this treatment, the pH condition, and the homogenization process (rotor/stator device and high-pressure homogenizer) were evaluated, determining emulsion properties such as phase separation, droplet size, and rheology. To analyze the creaming stability, aliquots of the emulsions prepared using the rotor/stator device and the high-pressure homogenizer were placed in graduated 10-mL (diameter = 15.5 mm, height = 65 mm) and 50-mL (internal diameter = 25 mm, height = 95 mm) test tubes. The larger cylinder used on fine emulsions' evaluation was chosen in order to reduce wall effects, since it is well known that confining walls exert an extra retardation effect on a spherical particle settling in a liquid (Chhabra et al. 2003). The stability of the emulsions against creaming was evaluated during 7 days of storage at room temperature (25°C) by visually monitoring the development of separate phases. The creaming index (CI%) of the emulsions was calculated as $\mathrm{CI}(\%) = (V_s / V_i) \times 100$, where V_i represents the initial volume of the emulsion and V_s the volume of the serum layer formed at the bottom of the tube (Keowmaneechai and McClements 2002). According to de Castro Santana et al. (2012), the alkaline emulsions showed lower kinetic stability, since collagen fibers have a lower net charge (ζ-potential) at higher pH values, decreasing the electrostatic stability process. Heat treatment slightly decreased the protein charge and significantly reduced the insoluble protein content, suggesting a decrease in the emulsifying properties of the collagen fiber.

3.2 Rheological Methods

The stability and rheological properties of emulsions are largely determined by the interactions between the droplets. The nature and strength of the interactions are, in turn, dependent on the structure and composition of the adsorbed layer at the oil–water interface. In food colloids, the stabilizing layer around emulsion droplets is compositionally and structurally complex. This obviously makes it difficult to disentangle the relationship between the colloidal interactions and the macroscopic emulsion properties (Dickinson 1998). One of the most important macroscopic properties of a food emulsion is its rheological properties. Texture and mouth-feel are of particular practical consequence, and the careful manipulation of emulsion rheology over a range of conditions is essential to provide a successful, marketable product. Food oil-in-water emulsions can range from mobile Newtonian liquids such as milk to highly non-Newtonian viscoelastic products such as mayonnaise. The rheological properties of emulsions are determined by many different factors, including the concentration of the dispersed phase, the emulsion stability, the temperature, and the viscosity (and composition) of the continuous phase. Manipulation of these various parameters can be influential in determining the rheological characteristics of an individual food emulsion product. Dickinson and Golding (1997a) proved that there was a close relationship between the rheological behavior of casein emulsions and protein concentrations using both steady-state viscometry and small-deformation oscillatory methods. That study demonstrated that small-deformation rheometry was a suitable tool for probing emulsion instability mechanisms that cannot readily be determined from visual or ultrasound creaming observations alone.

3.2.1 Small-Deformation Rheology

Interfacial rheology has been used to quantify the film strength of adsorbed emulsifier layers at fluid interfaces. There are two types of interfacial rheology: dilatational and shear. Dilatational techniques involve inducing a change in the interfacial area while simultaneously measuring the interfacial tension. Shear methods involve inducing shear in the film without a change in area; many different arrangements exist for this type of measurement in particular that allow measuring both mechanical (solid-like) and flow (liquid-like) properties of emulsions. Values of interfacial tension are very relevant to emulsion stability. However, for long-term emulsion stability to coalescence and hence phase separation, the strength of the interfacial film formed by a prosurface-active substance has been reported to be more important than its effect on interfacial tension (Opawale and Burgess 1998). Proteins have a polymeric and polyelectrolyte nature. Colloidal stabilization by protein adsorption is frequently the result of a combination of steric and electrostatic (electrosteric) repulsion. These forces originated for the adsorbed protein film impact interfacial rheology.

The rheological properties of oil-in-water (O/W) emulsions are generally controlled by varying the droplet volume fraction, the droplet size distribution, and, mainly, the interdroplet forces by additives. Following time-dependent changes on rheological properties of emulsions with a high oil-to-protein ratio (5.8–35), Dickinson and Golding reported in an early work (1997a) that the rheological behavior of caseinate emulsions was largely determined by the interactions between droplets and especially by the nature and strength of the interparticle attractive forces, which were dependent on the structure and composition of the adsorbed layer at the oil–water interface. The aim of that work was to provide experimental evidence in support of polymer theory. Their findings may be summarized as follows: The stability and rheology of emulsions made with NaCas depend on two main factors, the structure and composition of the adsorbed protein layer at the oil–water interface, and the state of self-assembly and aggregation of the protein in the aqueous phase. In emulsions containing protein at concentrations well below that required for monolayer saturation coverage, the system exhibits time-dependent bridging flocculation and coalescence. At protein contents around that required for saturation monolayer coverage, the system is Newtonian and unflocculated and is very stable toward creaming and coalescence. At higher protein contents, the creaming stability of the pseudo-plastic system is greatly reduced due to depletion flocculation of protein-coated droplets by unadsorbed submicellar caseinate. At even higher protein contents, there is partial restabilization of the flocculated emulsion in the form of a strong particle gel network. At low concentrations of protein, no flocculation was observed immediately after preparation, but the rheological behavior with time showed that there was a steady increase in droplet aggregation. The 35 vol.% *n*-tetradecane emulsions with protein concentration of 5 or 6% wt.% had not only an initial flocculation more extensive than emulsions with 1–4% wt.% protein but also showed a higher extent of structural rearrangement during aging. According to the authors, rearrangement to more closely packed structures would therefore tend to result in a greater reduction in apparent viscosity as more of the trapped continuous phase is released into the bulk. The rheological behavior of their systems demonstrated that the droplet networking effect was very relevant since with the casein systems even up to protein concentrations of 8 wt.%, there was relatively little contribution to the emulsion rheology from the viscoelastic properties of the continuous phase. So the structural mechanism influencing the rheology of the casein-rich emulsions can apparently be considered entirely attributable to inter-droplet depletion interactions.

Schokker and Dalgleish (1998) suggested that the functional properties of the casein layer at the oil–water interface could be related to the flocculation behavior. In the 20 vol.% soybean oil, 1 wt.% sodium caseinate aqueous solution emulsions they studied, the systems were more susceptible to shear-induced flocculation after preparation than during storage. However, no proteolysis products could be detected by SDS-PAGE, and also no exchange of protein between the droplet surface and the continuous phase was observed in the stored emulsions.

Stevenson et al. (1997) found that displacement of β-casein from the oil–water interface with Tween 20 was more difficult after storage at room temperature, presumably caused by rearrangements of molecules at the interface and an increase in hydrophobic interactions at the interface. A similar mechanism was proposed by Schokker and Dalgleish (2000) to interpret the decreased susceptibility with time of the emulsion to flocculate under shearing conditions.

Berli et al. (2002) proposed a theory to connect the interparticle forces and the rheology in order to manipulate the colloidal stability of emulsions. In agreement with Dickinson et al. (1997), they found that the rheological response of the emulsion was highly dependent on the concentration of caseinate. When the concentration of free proteins in the bulk solution was low, emulsions were Newtonian, while for higher caseinate concentrations, emulsions became shear-thinning. According to Berli et al. (2002), the variation of emulsion viscosity with protein concentration could not be explained by the variation in the suspending fluid viscosity. The strength of interparticle interactions and a genuine variation in the radius of the submicelle with formulation had a key role in rheological behavior.

In a recent work, Tipvarakarnkoon et al. (2010) studied the stabilizing effect of modified acacia gum on the stabilization of coconut O/W emulsions. For this purpose they prepared model substrates for coconut milk or coconut cream. Coconut milk is widely used as a food ingredient not only in Southeast Asia but also in Western countries. It is an O/W emulsion of varying fat contents, prepared from coconut meat by extraction. The native coconut proteins of the milk are not able to stabilize the emulsion sufficiently against creaming during storage. Therefore, various methods for coconut milk stabilization have been tested. On the one hand, different emulsifiers such as surfactants (Tween 20, SDS) as well as milk proteins (sodium caseinate) or whey protein isolate have been used (Jena and Das 2006; Tangsuphoom and Coupland 2008, 2009a, b). On the other hand, stabilizing carbohydrates such as coconut sugars or carboxymethylcellulose have been added (Jirapeangtong et al. 2008) alone or together with emulsifiers. Tipvarakarnkoon et al. (2010) used acacia gum since it has emulsifying as well as stabilizing abilities.

This native glycoprotein is well known and extensively used as an emulsifier and stabilizer in the food industry. The properties of native acacia gum, however, differ strongly in dependence of the raw material. Therefore, in this work it was modified by a maturation process that increased the emulsifying arabinogalactan protein share. All emulsions proved to be low-viscous and nearly Newtonian liquids. It can be concluded that the emulsion stability was mainly a result of the excellent emulsifying properties and not of an additional thickening effect of the gums. The modified acacia gums can be recommended as an emulsifier and stabilizer for application in different food products, preferably in low-viscous emulsions such as coconut milk drinks or other beverages.

Mixtures of surface-active substances are often used in many technological applications, including food and pharmaceutical industries, cosmetics, coating processes, and so forth. In many of these applications, protein–surfactant mixtures are used in the manufacture of processed dispersions. These dispersions contain two or more immiscible phases (aqueous, oil, and/or gas phases) in the form of foams and emulsions. In food systems, the interfacial layer often comprises both proteins and surfactants (mainly lipids and phospholipids). The optimum use of emulsifiers in food technological applications depends on our knowledge of their interfacial physicochemical characteristics, such as surface activity, amount adsorbed, the kinetics of film formation, structure, thickness, topography, ability to desorb (stability), lateral mobility, interactions between adsorbed molecules, ability to change conformation, interfacial rheological properties, etc. Maldonado-Valderrama and Rodríguez Patino (2010) published a review focused on the interfacial rheology of protein–surfactant mixtures, putting more emphasis on the interfacial dilatational rheology. According to the authors, literature reports has shown that interfacial rheology of protein–surfactant mixed films depends on the protein (random or globular), the surfactant (water-soluble or oil-soluble surfactant, ionic or nonionic), the interface (air–water or oil–water), the interfacial (protein/surfactant ratio) and bulk (i.e., pH, ionic strength, etc.) compositions, the method of formation of the interfacial film (by spreading or adsorption, either sequentially or simultaneously), the interactions (hydrophobic and/or electrostatic), and the displacement of protein by surfactant. Proteins and surfactants can form nano-scaled micelles and micro-scaled vesicular, crystalline, or other structures in the bulk phase prior to adsorption at the interface during dispersion formation. However, according to Maldonado-Valderrama and Rodríguez Patino (2010), little is known about the effect of these nano- and microstructures on the interfacial rheology and dispersion characteristics. Moreover, the relationship between the homogeneity/heterogeneity (i.e., aggregation, miscibility, phase separation, etc.) of a fluid interface and the macroscopic properties of the dispersion is only indirectly known from the rheological properties of the fluid interface. More research should be performed in order to establish quantitative correlations between the interfacial rheology (including the effect of pH, ionic strength, temperature, high pressure, addition of food reagents, protein modifications, etc.) and dispersion stability of pure and protein–surfactant mixed systems.

Murray (2011) has reviewed the scientific literature from 2002 to the present on the interfacial rheology of protein films, focusing on the implications for biological systems and in particular for food emulsions and foams. The areas covered include new methods of measurement; proteins and polysaccharides and protein/polysaccharide complexes; the effects of cross-linking within protein films and the origins of film viscoelasticity; proteins and low-molecular-weight surfactants; experimental and theoretical studies of the interfacial rheology and its relationship to emulsion and foam stability. According to Murray (2011), there has been a something of a resurgence of interest in these areas, resulting in a number of important advances that should aid further understanding and exploitation of proteins as surface-active agents and colloid stabilizers. As conclusions of his review, it may be mentioned that there have been a number of important advances in the study of the interfacial rheology of adsorbed protein films in the last 10 years or so. These have been most notable in instrumentation and measurement techniques, but also in an improved understanding of the behavior of proteins mixed with other surface-active agents or with molecules that can form complexes with the proteins that affect their adsorption. The diverse range of interfacial properties of proteins and other biosurfactants suggests that they could be more widely used as effective but biocompatible, biodegradable colloid stabilizers. Various new methods of measurement of the interfacial rheology of adsorbed protein together with improvements in the sensitivity and analysis of traditional methods have been described in detail in the review (Murray 2011).

3.2.2 *Viscosity*

Viscosity increase is a common strategy to enhance emulsion stability. This goal may be reached by adding compounds such as hydrocolloids or proteins. In this section, we'll describe some examples of products in which viscosity is a main property.

In the field of food additives, alcoholic cream liqueur emulsions are of special interest. The stability and shelf life of these emulsions depend on several factors, such as viscosity, volume particle size, temperature, pH, and ionic forces. The effect of caseinates is commonly underestimated; for example, the content of Ca^{++} and Na^{+} in caseinates influences the stability of alcoholic emulsions. A commercial use of caseinates in alcoholic cream liqueurs is as an emulsifying agent. Cream liqueurs of about 15% alcohol content can be prepared with an extended shelf life of several years. Medina-Torres et al. (2009) explored the effect of several kinds of commercial caseinates and their mixtures into the stability and shelf life of alcoholic cream liqueurs as estimated by rheological and volume particle size distribution behavior at different storage times and temperatures. Alcoholic emulsions were prepared with three different caseinate batches at 3% and 4% (w/v) protein content and 16.9% (v/v) ethanol content. Two storage temperatures (25°C and 40°C) and three storage times (0, 25, and 45 days) were used. The stability of the emulsion was followed visually by detecting coalescence. The preparation of emulsion for viscosity measurements was as follows: 70.3 g of a caseinate batch without alcohol (about 67.5 g of dry matter) was weighed in a 600-mL beaker (weight A, g) and cold water

(379.7 g) was slowly added using a rotor/stator system at 300 rpm for 20 min at 25°C. The mixture was stored for 12 h at room temperature to stabilize and to allow for lumps to be moistened. Then the beaker with its contents was placed in a water bath at 70°C for 30 min under occasional stirring to keep a homogeneous solution. After dissolving, the mixture was completed to the initial weight (A+450 g) to compensate for evaporated water, cooling the beaker under running water at 20°C for 4 h. The rheological measurements were performed on a strain-controlled rheometer using a double concentric geometry. The viscosity was measured at different storage times and temperatures (0, 15, 30, and 45 days, 25°C and 40°C, respectively). The protein content showed to be crucial to reduce coalescence. For all emulsions with 4% protein content, coalescence visually appeared at a higher alcohol content than for emulsions with 3% protein content. The volumetric diameter ($d_{4,3}$) was a function of storage time, and in all cases the coalescence increased. Viscosity also increased with storage time in all the blends. The viscosity was found to be directly related to the particle size of the emulsions prepared at different caseinate ratios. Casein emulsions with intermediate viscosity values were the most stable. Emulsions having low viscosity values are an advantage in the elaboration of liqueur creams because the formed micelles are more stable, avoiding the possible aggregation or precipitate formation. Although a small initial average droplet size in prolonged storage times does produce a short-term stability improvement, there was not yet a significant effect on long-term product stability. Finally, the results on microscopy suggested that the final structure plays an important role in the stability of the alcoholic emulsion with industrial interest in food emulsions, particularly in the elaboration of cream liqueurs.

Mango (*Mangifera indica* L.) is a tropical fruit in the plant family Anacadiaceae that originated in India and Southeast Asia. The mango pulp contains vitamins and bioactive compounds such as β-carotene and vitamins A, C, and E. Ripe mango pulp has β-carotene of 50% of total carotenoid and 2.0% (w/w) pectin, which is a soluble dietary fiber. In some countries, mango is mixed with milk to become cream stuff used in dessert for mango flavor, taste, and homogeneous texture. This mango cream is similar to custard, mayonnaise, and salad dressing, which are oil-in-water emulsions. Proteins and polysaccharides are commonly used together in oil-in-water food emulsions. Proteins are widely used as an emulsifier because they have an ability to adsorb at the oil–water interface and stabilize the oil droplets, while polysaccharides are usually added to increase the viscosity of emulsion (Dickinson 1995). Data in the literature indicated that both pectin and/or sodium caseinate had an effect on emulsion stability. To further investigate these ideas, Karunasawat and Anprung (2010) studied the effect of depolymerized mango pulp as a stabilizer in oil-in-water emulsion containing sodium caseinate. The objective of that paper was to investigate whether the degree of hydrolysis of mango pulp and the sodium caseinate concentration have effects on the oil-in-water emulsion stability in terms of average droplet size, viscosity, and creaming stability. The apparent viscosity of systems was measured by using a controlled stress rheometer equipped with cone and plate geometry. A cone diameter of 40 mm with a cone angle of 4° was used. The emulsion's viscosity was measured at room temperature (30°C ± 1°C) about

24 h after the emulsion was made. Increasing the degree of hydrolysis for pectin in mango pulp resulted in a decrease in the emulsion's viscosity. It can be seen that there was no change in the viscosity of emulsions over the experiment. This can be expected since the emulsions did not display flocculation or coalescence after a storage period of 1 day. The results reported by Karunasawat and Anprung (2010) indicated that the viscosity of the emulsions containing 0.5% or 1.0% (w/w) sodium caseinate was related to the average droplet sizes. It might be expected that nonflocculated samples behave as Newtonian fluids, while flocculated emulsions show pseudo-plastic behavior. As expected, the higher sodium caseinate concentration [>2.0% (w/w)] induced depletion flocculation in the emulsions. The entrapment of a certain amount of continuous phase in the flocculated structure caused an increase in the effective volume fraction of hydrodynamically interacting entities, which in turn increased the viscosity of aggregated emulsion systems according to the Dougherty–Krieger equation (McClements 1999). However, the flocculated structure was broken up with storage time, and some trapped continuous phase was released, consequently lowering the viscosity following the behavior described by Dickinson and Golding (1997a, b). For the emulsions containing sugar content of 46 mg of glucose/g fresh weight (DP 46E pectinase treated or DP 46), there was a decrease in viscosity with increasing shear rate, indicating shear-thinning (pseudoplastic) behavior of the emulsions containing DP 46E or DP 46. The emulsion with the best stability was that made from depolymerized mango pulp with reducing sugar content of 60 mg of glucose/g fresh weight (DP 60) and 2% (w/w) sodium caseinate. In addition, it was found that DP 60 could be used as an alternative stabilizer for oil-in-water food emulsions.

Alginate is a linear copolymer with homopolymeric blocks of (1–4)-linked β-D-mannuronate (M) and its C-5 epimer α-L-guluronate (G) residues, respectively, covalently linked together in different sequences or blocks. Alginate has a number of free hydroxyl and carboxyl groups distributed along the backbone; therefore, it may be functionalized by chemical modification. By forming alginate derivatives through functionalizing available hydroxyl and carboxyl groups, properties such as solubility and hydrophobicity and physicochemical and biological characteristics may be modified (Yang et al. 2011). Alginates are widely used as a gelling agent for thickening foods and cosmetics. Yang et al. (2012) modified alginate with the aim of using it as an emulsifier in emulsions. The stability of their systems was followed by visual observation, droplet size, microstructure, and viscosity. Viscosity measurements were conducted in a rheometer using a cone-and-plate geometry, with a cone angle of 1° and a diameter of 60 mm. The samples were introduced onto the plate for 5 min to eliminate residual shear history and then experiments were carried out immediately. The measuring device was equipped with a temperature unit that gave good temperature control (25°C ± 0.05°C) over an extended time. Their studies showed that the emulsions containing 1.0 wt.% sodium alginate and 0.3 and 0.5 wt.% dodecanol alginate were unstable, whereas the emulsions containing 0.8–1.2 wt.% dodecanol alginate presented better stability during storage.

It is well known that the polymer–surfactant complexation in colloidal formulations can alter their stability and rheology. Studies show that the stabilization of emulsions by using a combination of surfactant and polyelectrolytes (Stamkulov et al. 2009) or solid nanoparticles (Binks et al. 2007) can synergistically enhance their stability. Nambam and Philip (2012) investigated the competitive adsorption of polymer and surfactant at an oil–water interface by measuring the hydrodynamic diameter, ζ-potential, microstructure, and rheology. The polymer selected was a statistical copolymer of polyvinyl alcohol and vinyl acetate copolymer (PVA–VAc), and the emulsion was an oil-in-water system with an average droplet diameter of 200 nm. The oil consisted of oleic acid–stabilized Fe_3O_4 nanoparticles of size 10 nm and the surfactant sodium dodecyl sulfate (SDS), cetyltrimethylammonium bromide (CTAB), and nonylphenol ethoxylate (NP9). The questions they tried to address from those studies were the following: What was the equilibrium situation of polymer–surfactant complexation when both the polymer and surfactant have abilities to complex with each other and at the same time they can also adsorb at the droplet interface independently? What is the consequence of such complexation on the stability (agglomeration) of emulsions? What are the conditions to minimize flocculation of emulsion droplets? The enhanced viscosity upon the addition of ionic surfactant into polymers confirms the strong interaction between them. Their studies show that lower-molecular-weight polymers with suitable ionic surfactants can synergistically enhance the stability of formulations, while longer-chain polymers induce bridging flocculation. Their results were useful for preparing oil-in-water formulations with long-term stability.

3.3 Ultrasound Profiling

In a pioneering work, Dickinson and Golding (1997b) followed the stability of emulsions with a high oil-to-protein ratio (5.8–35) by the technique of ultrasonic velocity scanning. The bases of the technique are as follows: The velocity of sound in a dispersion is related to the physical properties of dispersed and continuous phases and the relative quantities of each phase. When a planar sound wave is incident on a spherical particle, the wave is scattered; that is, a proportion of the energy is removed from the forward direction of the wave and is transmitted in other directions. The phase changes in the forward component of the wave manifest themselves as a change in the apparent velocity of the sound wave. Thus, the velocity becomes a function of parameters such as particle size and the differences in physical properties between the two phases. Creaming of an emulsion occurs when the dispersed phase is less dense than the continuous phase and so drifts to the top of the sample, forming a cream layer. This method allows one to detect the creaming destabilization of emulsion systems by recording the profiles of the dispersed-phase concentration as a function of heights at different times. These measurements provide the means for stability studies through the use of an equation to relate the ultrasound velocity to the dispersed-phase concentration, that is, the volume fraction. The approach in each case depends on the systems under study and is usually closely associated with the Urick equation (Pinfield et al. 1995). The model systems studied by Dickinson and Golding (1997b) were formulated with NaCas (1–6 wt.%) as the sole emulsifier, *n*-tetradecane as the fat phase (35 vol.%), and a phosphate buffer as the aqueous phase. Their results showed that creaming and flocculation kinetics had a complex dependence on caseinate content. The most stable caseinate emulsions with respect to creaming were those for which there was sufficient protein present to give excellent surface coverage and associated steric/electrostatic stabilization, but where the concentration of unbound protein was still low (below the critical flocculation concentration) so that there was no depletion flocculation.

In an emulsion, the volume fraction of each phase is defined as the radio of the volume of each phase over the total volume. The dispersed phase's volume fraction and droplet size distribution of emulsions determine many of their most important physicochemical properties, including their stability, rheology, and appearance. Usually, a high volume fraction for the dispersed phase induces an opaque emulsion and prevents the utilization of light to analyze it. A way to circumvent this difficulty of analyzing by light is the use of ultrasonics as an inquiry method. Polysaccharides may interact with protein adsorbed at the interface. Depending on the size and charge of the molecules, the concentration of polysaccharide present, and the environmental conditions of the solution (pH and ionic strength), the protein–polysaccharide interactions can (1) improve the stability of the emulsion (may have an associative interaction), (2) lead to destabilization by bridging flocculation, where a long chain polymer is present in small concentrations and adsorbs onto more than one colloidal particle, or depletion flocculation, where the intercolloidal region becomes depleted of polymer, creating a polymer concentration gradient, hence an osmotic pressure difference, which draws the particles closer to one another, or (3) last, cause gelation. Understanding the assemblies occurring at the interface and the dynamics of the interactions is fundamental for engineering food emulsions. Ultrasonic spectroscopy has been employed to characterize dispersed systems such as emulsions and milk and to study protein–polysaccharide interactions (Dwyer et al. 2005; Corredig et al. 2004a; Dalgleish et al. 2005). The ultrasonic waves have the ability to propagate through optically opaque materials and therefore are highly suited for use in concentrated colloidal systems. The parameters obtained are sensitive to the molecular organization and molecular interactions in the samples. Ultrasonic spectroscopy is therefore noninvasive and can determine changes in physicochemical rearrangements in colloidal systems. In studies focusing on structure development, the stages preceding aggregation and the sol-to-gel transition are of extreme importance, as these moments are critical for the physical and sensorial characteristics of the final gel. With conventional techniques such as traditional light scattering or rheology, information on the liquid-to-solid transition state might be lost. Ultrasound spectroscopy also shows great potential for the study of in situ changes in emulsion systems. This technique has been successfully employed to study droplet–droplet interactions in situ to distinguish differences between bridging flocculation and depletion flocculation in whey protein–stabilized emulsions (Gancz et al. 2005).

High-methoxyl pectins (HMP) are negatively charged polysaccharides composed of a backbone of methyl-esterified galacturonic acid (>50% substituted) with branched regions containing arabinose, galactose, and xylose. Changes in charge and charge distribution on the pectin molecule strongly affect the interactions with proteins. Liu et al. (2007a) described the application of high-resolution ultrasonic spectroscopy on the study of the acid-induced aggregation of sodium caseinate emulsions in the presence of HMP. The effect of pH as well as the concentration and charge (degree of esterification, DE) of HMP on the structural changes were observed by measuring ultrasonic velocity and attenuation. They found that during acidification, caused by the addition of glucono-δ-lactone, there were small changes in the overall ultrasonic velocity. Although small, it was possible to relate these changes to the structural changes in the emulsion. The values of ultrasonic attenuation decreased at high pH with increasing amount of HMP, indicating changes in the flocculation state of the oil droplets caused by depletion forces. During acidification at pH 5.4, emulsions containing HMP showed a steep increase in the ultrasonic attenuation, and this pH corresponds to the pH of association of HMP with the casein-covered oil droplets. The adsorption of HMP onto the interface causes a rearrangement of the oil droplets, and the emulsions containing sufficient amounts of HMP no longer gel at acidic pH. This research demonstrated for the first time that ultrasonic spectroscopy can be employed for in situ monitoring and analysis of acid-induced destabilization of food emulsions.

One of the major disadvantages of applying the ultrasonic technique to emulsions is that a considerable amount of data about the thermophysical properties of the component is needed to interpret ultrasonic spectra, such as adiabatic compressibility, density, viscosity, conductivity, specific heat capacity, and cubical expansion coefficient. Values of these properties for a variety of liquids have been compiled. Nevertheless, all of these thermophysical properties are temperature-dependent, which means that the ultrasonic properties of emulsions vary with temperature. Chanamai et al. (1998) investigated the influence of temperature on the ultrasonic velocity and attenuation coefficient of oil-in-water emulsions with different compositions. They also explored the implications of the temperature dependence of the ultrasonic properties of emulsions for the utilization of ultrasonic techniques for characterizing emulsions. Their results indicated that it is important to either carefully control or measure the temperature during an ultrasonic analysis. The measured ultrasonic properties of the emulsions were in reasonable agreement with those predicted by ultrasonic scattering theory across most of the temperature range. Nevertheless, there were some significant disagreements, especially at high temperatures, which thus highlight the need for researchers to accurately measure and compile the temperature dependence of the thermophysical and ultrasonic properties of a wide variety of different liquids.

A wide variety of experimental techniques have been developed in the past 30–40 years to characterize the droplets in emulsions, including electron and light microscopy, dynamic and static light scattering, neutron scattering, and electrical conductivity measurements. Most of these methods cause some amount of disruption, during either the analysis or sample preparation stages. This makes it difficult to accurately analyze in detail the changes occurring during colloidal destabilization, namely, sol–gel transitions or phase separations, as these mechanisms are dependent on the volume fraction or minimal shear disturbance. In addition, most of these techniques are either destructive or only suitable for application to dilute optically transparent systems, whereas most emulsions of practical importance are concentrated and optically opaque. It was clear at that time that there was a need for analytical techniques to provide information about the physicochemical properties of concentrated emulsions. Thus, for this purpose, new techniques such as NMR and ultrasonic spectroscopy were developed. McClements and Coupland (1996) proved the fitness of ultrasound spectroscopy for measuring droplet size distributions in concentrated emulsions in situ. The ultrasonic velocity and attenuation coefficient of an emulsion are measured over a range of frequencies, and then the multiple-scattering theory is used to convert these measurements into a droplet size distribution. The theory they selected assumes that droplets are spherical, much smaller than the ultrasonic wavelength, and do not physically interact with their neighbors and that higher-order scattering terms are negligible. According to McClements and Coupland (1996), the first two assumptions are valid for most emulsions. Droplets only become nonspherical at high-volume fractions, in high-shear fields, or when they crystallize. The majority of emulsion droplets of importance are between 0.1 and 10 μm, whereas the ultrasonic wavelength is greater than 150 μm at frequencies of 10 MHz and less. The latter two assumptions may be violated in concentrated emulsions or in systems where the droplets flocculate, and can lead to significant deviations between experimental measurements and theoretical predictions. Despite these limitations, the ultrasonic theory presented by McClements and Coupland (1996) has been shown to be applicable to a number of oil-in-water emulsions up to fairly high-volume fractions (20–30%). One advantage of this technique is that it can easily be adapted for online measurements, which could be of considerable importance for many manufacturing processes. One of the major limitations of the ultrasonic technique is that it cannot be used to study emulsions that contain small gas bubbles. This is because the gas bubbles scatter the ultrasound so effectively (even at very low concentrations) that the ultrasonic signal may be completely attenuated. Ultrasound also has limited application to very dilute emulsions (<0.5 wt.%) because the change in ultrasonic properties with droplet size becomes of the same order as the experimental error. In summary, ultrasonic spectroscopy is most useful for studying concentrated or optically opaque materials, or when online measurements are required. For dilute emulsions (<1%), light scattering techniques are usually preferable.

Gülseren and Corredig (2011) reported that there are four major mechanisms that contribute, generally additively, to the magnitude of attenuation in colloidal systems, and they are dependent on the material properties: *Intrinsic attenuation* is a material property and is observed in continuous phases as well as in dispersed phases; *visco-inertial losses* are due to the density differences between the dispersed and the continuous phases; *thermal losses* are due to the temperature gradient between the continuous and dispersed phases that is generated by the ultrasonic compression-decompression cycles; and *scattering losses* take place as an ultrasound wave is refracted, diffracted, and reflected by the dispersed phase particles. Increasing ultrasonic frequencies (i.e., decreasing wavelengths) enable us to investigate increasingly smaller size scales, and scattering attenuation becomes comparatively larger with frequency. According to the authors, the primary loss mechanism in food emulsions of a few μm in diameter is thermal losses, especially at frequencies lower than 100 MHz. In agreement with McClements and Coupland (1996), they concluded that based on the material properties, the measured frequency dependence of attenuation can be exploited to derive particle size distributions.

3.4 Electroacoustic Spectroscopy: ζ-Potential

Over the past two decades, electroacoustic (EA) techniques have emerged as a powerful means of monitoring droplet charge (ζ-potential) in colloidal systems. As happens with ultrasonic spectroscopy, electroacoustic spectroscopy is a nondestructive technique. It also has the potential to analyze concentrated colloids or emulsions online and in situ and does not suffer from opacity of samples, a major problem for most light scattering techniques. Thus, the major advantage of the EA technique over more conventional microelectrophoretic techniques based on light scattering is that it is capable of analyzing emulsions with high droplet concentrations (<50%) without any sample dilution. EA spectroscopy can also be used to provide information about the droplet size distribution of emulsions; however, the droplet size range is usually rather limited (0.1–10 μm). Nevertheless, there are some limitations of the EA technique for certain applications. For example, the droplets must have an electrical charge, there must be a significant density contrast ($\Delta\rho > 2\%$) between the droplets and the surrounding liquid, and the viscosity of the continuous phase must be known at the measurement frequency (which is not always the same as that measured in a conventional viscometer) (Cho and McClements 2007).

Electroacoustic spectroscopy couples ultrasonic and electric fields. Currently, there are two separate approaches to electroacoustics: the electrokinetic sonic amplitude (ESA) approach and the colloidal vibration current (CVI) approach. In the ESA method, a radio frequency signal is applied to an emulsion and the resulting acoustic signal generated by the oscillating particles is recorded. Cho and McClements (2007) demonstrated that the ζ-potential of oil droplets measured in situ were consistent with similar measurements based on light scattering (i.e., highly diluted emulsions). Both techniques were able to monitor the adsorption of pectin onto the surfaces of β-lactoglobulin-coated droplets as a function of pectin concentration and pH. The major advantage of the EA technique was that it could be carried out in situ without disturbing the equilibrium between adsorbed and nonadsorbed polyelectrolyte. Nevertheless, the good agreement between the ζ-potential values determined by the EA and microelectrophoresis techniques suggested that emulsion dilution did not cause an appreciable change in polysaccharide partitioning for the system used in their study. In the CVI approach, the sample is subjected to a longitudinal ultrasonic pulse. The vibrations generated by the planar ultrasonic wave compressions, in turn, cause the relative motion of particles to the surrounding bulk phase. This motion perturbs the electrical double layers of the surrounding colloidal particles and causes an electrical response (CVI). The electrophoretic mobility and ζ-potential can be derived from the CVI signal, the ultrasonic properties of the particles, and the density contrast between the continuous and dispersed phases. Beattie and Djerdjev (2000) showed that the changes occurring during Ostwald ripening in emulsions could be followed using electroacoustics. Using an ESA-based method, Mun et al. (2008) investigated the possibility of using polysaccharide coatings to improve the freeze–thaw and freeze–dry stability of protein-coated lipid droplets. In that study, emulsions were diluted to a droplet concentration of approximately 0.006 wt.% oil using buffer solution prior to analysis. Diluted emulsions were then injected into the measurement chamber of a particle electrophoresis instrument and the ζ-potential was determined by measuring the direction and velocity that the droplets moved in the applied electric field. The ζ-potential measurements were reported as the average and standard deviation of measurements made on two freshly prepared samples, with five readings made per sample. Their results show that forming a polysaccharide layer around the protein-coated lipid droplets in an oil-in-water emulsion cannot prevent destabilization during freeze–thaw cycling or freeze–drying in the absence of maltodextrin. Nevertheless, a polysaccharide coating can appreciably decrease the concentration of maltodextrin required to create emulsions that are stable to freeze–thawing or freeze–drying. This may have important consequences for the development of reduced-sugar or reduced-calorie frozen or powdered emulsion products. Gülseren and Corredig (2011) used the CVI approach to describe oil-in-water emulsions stabilized by sodium caseinate. The systems were employed as a model to observe the changes in acoustic and electroacoustic parameters during different types of flocculation caused by changes in pH or by the presence of polymers in the continuous phase. The electroacoustic and ultrasonic properties of soy oil-in-water emulsions were determined for sodium caseinate–stabilized emulsions under conditions known to cause destabilization. Ultrasonic attenuation and electrophoretic mobility (ζ-potential) could clearly follow the changes occurring in the emulsion droplets, under minimal sample disruption. This is critical for these systems in a very fragile, metastable state. Destabilization was generally characterized by a reduction in attenuation, a decrease in electrophoretic mobility, and an increase in mean particle size. Electroacoustics seemed to be very sensitive to monitoring structural rearrangements. In detail, the emulsions were stable to the addition of high-methoxyl pectin (HMP) up to 0.1% HMP. The addition of free sodium caseinate induced depletion flocculation, causing a decrease in the attenuation and electrophoretic mobility measured. The presence of HMP limited depletion interactions. Acidification of the emulsion droplets resulted in a clear sol–gel transition, as shown by a steep increase in the particle size and a decrease in attenuation. Again, destabilization was limited by HMP addition. It was concluded that ultrasonics and electroacoustics are suitable techniques to understand the details of the destabilization processes occurring in food emulsions, measured in situ since different types of destabilizations were monitored and compared with other well-known techniques such as light scattering measurements.

The fundamental measurements that have been employed in the electroacoustic and ultrasonic analyses include electroacoustic parameters such as surface charge (i.e., ζ-potential) and electrophoretic mobility of the colloidal particles or ultrasonic parameters such as velocity and attenuation. All of these parameters have been used to investigate real-time changes in emulsions and colloids. Other good examples of applications of these techniques may be found in Corredig et al. (2004b), Dukhin et al. (2010), and Gülseren et al. (2010).

3.5 Measurement of Surface Concentration

Physical properties of oil-in-water emulsions stabilized by milk proteins are determined largely by the nature of the adsorbed layer at the surface of the dispersed droplets. The creaming behavior and droplet flocculation are also sensitive to the concentration of nonadsorbed casein. It was reported that in casein-stabilized emulsions, at protein contents around that required for saturation monolayer coverage, the systems are very stable toward creaming and coalescence (Dickinson 1999). Thus, measurements of protein surface concentration are very relevant since they are closely related to the stability of emulsions. Although the interface plays a critical role, the interfacial protein concentration cannot be studied directly. Kalnin et al. (2004) presented a new method for the determination of particle surface coverage by NaCas based on density measurements. In this respect, easy-to-apply techniques are well suited to study colloids of biological, pharmaceutical, medical, or industrial interest. In their experiments, emulsions containing 20% fat were stabilized with 1–4 wt.% NaCas, which leads to a different size distribution using the same homogenization pressure. The total amount of NaCas, $[\mathrm{NaCas}]_{\mathrm{tot}}$, partitions into the amount of NaCas at the interface, $[\mathrm{NaCas}]_{\mathrm{lip}}$, and the free NaCas, $[\mathrm{NaCas}]_{\mathrm{aq}}$, according to

$$[\mathrm{NaCas}]_{\mathrm{tot}} = [\mathrm{NaCas}]_{\mathrm{lip}} + [\mathrm{NaCas}]_{\mathrm{aq}}$$

Their results showed that for small amounts of $[\mathrm{NaCas}]_{\mathrm{tot}}$, the concentration seems to limit the fat globule size in the emulsion, whereas for higher concentrations, the homogenization pressure seems to be the limiting factor, since the globule size distribution does not change significantly at $[\mathrm{NaCas}]_{\mathrm{tot}} > 3$ wt.%. This means that there will be a partition expressed as

$$K = [\mathrm{NaCas}]_{\mathrm{lip}} / [\mathrm{NaCas}]_{\mathrm{aq}}$$

between the aqueous phase and the interface depending on the surface area of the interface. They supposed for the further interpretation that the gravitational force that was apply to reinforce the creaming procedure was too weak to deplete the fat globules of their $[\mathrm{NaCas}]_{\mathrm{lip}}$ and only separates the aqueous phase "as is" without changing this partition. Therefore, $[\mathrm{NaCas}]_{\mathrm{lip}}$ was determined by density measurements of the aqueous phase resulting from the centrifuged emulsion. Provided that the partition coefficient K is not affected by the centrifugation step, Kalnin et al.'s (2004) new method based on density measurements allows the quantitative characterization of NaCas content in the aqueous phase of oil-in-water emulsions. Once K is known and the mean size of the emulsion droplets has been measured, the interfacial composition can be determined after measuring the "excess" concentration in the aqueous phase.

3.6 Microscopic Analysis

3.6.1 Transmission Electron Microscopy (TEM)

An important part of our understanding of materials is derived from studies of their behavior under changing conditions of temperature, stress, environment, etc. The basis of many of these studies is the direct relationship between the microstructure and the properties of materials, whereby an understanding and characterization of the former lead to an explanation of the latter. Transmission electron microscopy has proved to be the most successful instrumental technique for this purpose by providing structural, morphological, and compositional information from small volumes of thin-foil specimens.

O/W emulsions or W/O creams may be complex multicomponent preparations. They are combinations of a number of surfactants, polymers, and other additives. Mixed emulsifiers are frequently used in O/W systems; these are the combinations of an ionic or nonionic surfactant with fatty amphiphiles, such as fatty alcohols, fatty acids, or monoglycerides. Berg et al. (2004) developed methods to study W/O emulsions that make it possible to study the structure of complex multicomponent mixtures while largely preserving their structure and to predict their stability. The samples were cryoimmobilized using the jet freezing technique with liquid propane. In this method, the redistribution of ingredients or phase transitions is avoided in the products under investigation by using cooling rates of approximately 30,000 K/s. After cryoimmobilization, the sample was freeze-fractured by using a special fracture holder with specimen sandwiches. It was placed on the precooled specimen table that allows fracture of the cryoimmobilized sample to form two complementary fracture areas. Water was sublimated at 173 K and a pressure of 2×10^{-7} hPa to expose surface structures to a depth of approximately 90 nm. An anticontaminator directly above the sample prevented the contamination of the fresh fracture face by water recondensation. The internal structures exposed by fracturing and etching were replicated by means of a thin platinum/carbon replica. First, 2 nm of platinum/carbon was evaporated onto the surface at an elevation angle of 45°. This gave rise to electron-dense and less electron-dense areas (shadows) according to the surface topography in the replica that contribute to the contrast in the TEM. The replica prepared in this manner was then stabilized by additional evaporation of about 20 nm of carbon onto the surface at an angle of 90°. The respective shadow-casting film thickness was adjusted using an oscillating quartz crystal thickness monitor. After shadowing, the replica was floated onto organic solvent (ethanol or acetone) followed by washing in double-distilled water to remove adhering residues.

Then the samples were placed on a Cu-grid. A complex commercial W/O emulsion (25 components) with a volume fraction of $\varphi = 0.67$ together with 12 model emulsions with a variable volume fraction φ (0.5–0.86) were studied by TEM. These studies showed a transition from a particle solution below a critical φ_c to a close-packed network above that φ_c.

With the aim of clarifying the role of the mixed emulsifier in the structure formation and water binding mode in the case of O/W creams prepared with different surfactants, Kónya et al. (2007) analyzed the swelling behavior of mixed emulsifiers by means of direct investigation methods such as transmission electron microscopy (TEM) and X-ray diffraction. A series of ternary surfactant/CSA/water systems were prepared with a constant ratio of surfactant to fatty alcohol of 1:4 and increasing water contents 25%, 40%, 60%, 70% w/w. The results revealed that the investigated creams had different structures from those mentioned in the literature. The surfactant of the mixed emulsifier formed micelles, instead of mixed bilayers with the fatty amphiphile.

Cryo-TEM is a widely used technique for the characterization of a variety of self-assembled nanostructures. In particular, valuable information on the interparticle structure of cubosomes, hexosomes, micellar cubosomes, and emulsified microemulsion droplets has been obtained when using this method as a complement to scattering techniques (SAXS and SANS) and dynamic light scattering (DLS) (Yaghmur and Glatter 2009). With respect to the significant developments in the field of electron microscopy regarding the characterization of nanostructured aqueous dispersions, recent studies have suggested the use of tilt-angle cryo-TEM and cryo-FESEM for providing valuable insight into both the interparticle confined nanostructures and the 3D imaging of dispersed particles. The tilt-angle cryo-TEM paves the way as a powerful tool for characterizing the internal nanostructure in various nanostructured dispersions. Controlling the tilt angle allows a differentiation between different nanostructured dispersed particles (cubosomes, micellar cubosomes, and hexosomes) and also between the two different interiors in observed cubosome particles. This method could also be useful for distinguishing between different internal nano-objects within a dispersion. Boyd et al. (2007) introduced the cryo-FESEM technique for directly investigating the 3D and surface structures of both nondispersed liquid-crystalline phases and the dispersed cubosome and hexosome particles. In that publication, they pioneered the characterization of the 3D morphology of cubosome and hexosome particles. They demonstrated that the 3D cubosome structure enclosing aqueous water channels agreed well with the proposed mathematical models using a nodal surface representation.

Using microemulsions (ME) as delivery vehicles requires understanding whether water-insoluble molecules are delivered by an interaction of the ME system with the dietary mixed micelles (DMM) in the small intestine to give new mixed micelles, or by alternate paths. To answer this question, Rosner et al. (2010) diluted DMM and ME systems at various weight ratios. Based on DLS and cryo-TEM, they found a decrease in the average droplet diameter and an increase in the droplet density per unit area compared to pure ME systems. Their results showed that DMM and ME interacted to create ME–DMM mixed micelles, providing a potential pathway for delivering solubilized molecules.

It is well known that the bioavailability of protein and peptide drugs after oral administration is very low because of their instability in the gastrointestinal tract and low permeability through the intestinal mucosa. Numerous methods have been employed to solve this problem, including chemical modification, addition of enzyme inhibitor or absorption enhancer, conjugation with receptor-recognizable ligand, using particulate delivery carrier systems, and so on. Nanoparticle is the most commonly used method because it can protect the drug from the attack of gastric acid or pancreatic enzyme and then promote drug absorption by endocytosis or other mechanisms. Solid lipid nanoparticle (SLN), composed of physiological compatible lipid, has been used successfully to improve the bioavailability of protein and peptide. Given that SLN has a solid structure with strong hydrophobicity, gastric acid and protease enzyme cannot penetrate it and the drug load is protected greatly. Compared with poly(lactic-co-glycolic acid) (PLGA) nanoparticle, the degradation products of SLN are weaker acids that have little effect on the stability of protein and peptides. However, due to their hydrophilic nature, most proteins are poorly encapsulated into the hydrophobic matrix of SLN, tending to partition in the water phase during the preparation process. Therefore, the drug-loading capacities are extremely low. Yang et al. (2010) developed protein-loaded SLN with high entrapment efficiency (EE). For this purpose, hydrogel was selected as the carrier for protein, due to its hydrophilic nature, to achieve high entrapment of protein and because hydrogen bonds may be formed between hydrogel and protein. If hydrogel is implanted into SLN, its compatibility with protein can be utilized to improve the EE of SLN. The particle sizes and ζ-potential were characterized by dynamic light scattering and electrophoretic light scattering. Transmission electron microscopy (TEM) was employed to investigate the structure of this gel-core-SLN. The gel-core-SLN was successfully prepared and the particle size was 305.2 nm, with a ζ-potential of -17.15 mV. Observations by TEM confirmed that most solidified hydrogel particles were dispersed in the center of the gel-core-SLN in the form of a single core, which effectively prevented the diffusion of proteins to the external water phase during the preparation process.

Traditionally, synthetic surfactants or surface-active polymers, which tend to be adsorbed onto the interface and thereby reduce the interfacial free energy, were employed as emulsifiers in the preparation of stable emulsions. With increasing legal and consumer requirements such as nontoxic, biocompatible, mild to skin, high ecological acceptability, attractive price-to-performance ratio, and degrees of freedom in selecting and designing, these classic emulsifiers are becoming more and more limited. On this background, Pickering emulsions, which are emulsions stabilized solely by fine solid particles instead of surfactants, have attracted more and more interest due to their special properties. Pickering emulsions show favorable properties in comparison to classic surfactant-stabilized emulsions. The highly enhanced stabilization against coalescence and Oswald ripening makes it possible to conserve the droplets under a high concentration of dispersed phase, and they are even allowed to dry and redisperse. In addition, they are usually insensitive to changes in chemical parameters such as pH, oil composition, adding electrolytes, etc. Furthermore, additional stabilization can be achieved when the particles aggregate and form a three-dimensional network in the continuous phase, in which the droplets are captured in the array of particles. A wide variety of solid particles, including organic particles (such as polymer latex and polymer micelle) and inorganic particles (e.g., silica, hydroxides, and clay particles), have been used for the stabilization of emulsions. Chen et al. (2011) performed rheological studies using differently modified dispersible colloidal Boehmite alumina nanoparticles. Differential scanning calorimetry (DSC) results, transmission electron microscopy (TEM) images, and optical microscopy images of these emulsion systems were additionally analyzed to reveal the emulsions' type and their microstructure. Pickering emulsion stabilized by moderately hydrophobic particles exhibited an inhomogeneous structure, relatively large yield stresses, and thixotropic flow behavior, indicating a formation of a three-dimensional network. An emulsion stabilized by rather hydrophobic particles with a contact angle around 90° was revealed to be an oil-in-water-in-oil multiple emulsion. The emulsion was homogeneous and showed thixotropy, indicating the presence of a three-dimensional network. However, a phase separation slightly occurred in the storage time evaluated, meaning that the stability of this emulsion needs to be improved.

3.6.2 CLSM

Light microscopy is a well-developed and increasingly used technique for studying the microstructure of food systems in relation to their physical properties and processing behavior. Good-quality, high-resolution images of the internal structures of foods can only be obtained from thin sections of the sample. Procedures that applied substantial shear and compressive forces may destroy or damage structural elements, and sectioning is time-consuming and involves chemical processing steps that may introduce artifacts and make image interpretation difficult. Confocal laser-scanning microscopy (CLSM) overcomes these problems. In this instrument, image formation does not depend on transmitting light through the specimen, and, therefore, for the first time, bulk specimens can be used in light microscopy. The instrument uses a focused scanning laser to illuminate a subsurface layer of the specimen in such a way that information from this focal plane passes back through the specimen and is projected onto a pinhole (confocal aperture) in front of a detector. Only a focal plane image is produced, which is an optical slice of the structure at some preselected depth within the sample. By moving the specimen up and down relative to the focused laser light, a large number of consecutive optical sections with improved lateral resolution (compared with conventional light microscopy) can be obtained with a minimum of sample preparation. CLSM has been used in food science since the 1980s. Several reviews discussing the application of this technique in microstructural studies of food products have been reported in the literature (Heertje et al. 1987; Blonk and Van Aalst 1993; Marangoni and Hartel 1998). These reviews have shown the advantages of using CLSM over conventional techniques for studying the relationships among the composition, processing, and final properties of these products (Herrera and Hartel 2001).

CLSM has also been used to successfully describe emulsions' structures. As the fatty acid esters of propylene glycol are widely used in the manufacturing of food and cosmetic emulsions, Macierzanka and Szeląg (2006) studied microstructural properties of emulsions prepared with these compounds. Mono-diester products of the esterifications of propylene glycol with fatty acids were used as stabilizers of W/O emulsions. The emulsifiers also contained some amounts of unreacted fatty acids and propylene glycol, as well as zinc carboxylates, the compounds that were used in order to accelerate the esterification. An interfacial crystallization and the formation of the continuous oil-phase crystalline network of the acyl propylene glycols produced very lasting barriers against the sedimentation and coalescence of water droplets. The relative concentrations of monoacyl propylene glycol/zinc fatty acid carboxylate, the two main emulsifying components of the emulsifiers used, influenced the microstructure and, as a consequence, the rheological properties of the finally obtained W/O emulsions. By increasing the concentration of zinc carboxylates, an easier flow of W/O emulsions can be obtained. This is because a different structure of the emulsions is formed for various monoacyl propylene glycol/zinc carboxylate proportions. This work showed the importance of the microstructure in stability behavior.

The constituents of foods, or the foodstuffs themselves, are often not regarded as materials in the conventional sense. However, in reality, they are invariably highly complex composites that require thorough structural characterization if their physical behavior is to be fully understood. Only rarely is the actual evolution of the structure assessed dynamically during such experiments, as the majority of characterization techniques for these soft-solid materials require that the structure is set in some way prior to examination (i.e., freeze-drying to remove water). Conversely, the use of CLSM generally allows the materials to be examined in their natural state (i.e., hydrated) in such a way that structural evolution due to mechanical or thermal perturbation can be readily followed. Because of this, CLSM is a powerful tool to study gels. Plucknett et al. (2001) examined mixed biopolymer gels by CLSM, with the application of dynamic mechanical deformation during visualization. Specifically, they focused on the large strain deformation and failure behavior of two immiscible phase-separated composite biopolymer gel systems, namely, gelatin/maltodextrin and gelatin/agarose. Their results showed that the interfacial fracture energy of the mixed biopolymer system plays a significant role in determining the failure mode. For a relatively weak interface, such as for the gelatin/maltodextrin system (when gelatin is the continuous phase), debonding of the interface occurs at relatively low strains, resulting in a pseudo-ductile stress/strain response. For the gelatin/agarose system (again when gelatin is continuous), where the interfacial fracture energy is an order of magnitude higher, only limited debonding occurs and the material is nominally linear to failure. Direct evidence for debonding, in both cases, was obtained by conducting tension and compression tests in situ on the CLSM.

Other gels studied by CLSM were those formulated with whey proteins. Whey proteins, as important nutritional and functional food ingredients, have been extensively used in various food applications, such as sport beverages, meat replacements, baked products, salad dressings, ice creams, artificial coffee creams, soups, and various dairy products (Dissanayeke and Vasiljevic 2009). These proteins are excellent foaming and emulsifying agents and easily diffuse to the newly formed water–oil interface (e.g., during the emulsification), unfold and reorient in ways that greatly lower interfacial tension, and subsequently form cohesive and viscoelastic films mainly by disulfide bonds and hydrophobic interactions. Manoi and Rizvi (2008) showed that the conformational structure and functionalities of whey protein concentrate (WPC) could be modified through partial denaturation by means of combined treatments of highly acid treatment ($pH < 3.0$) with heat, shear, and supercritical carbon dioxide (SC-CO_2) injection during a supercritical fluid extrusion process (SCFX) in the presence of optimum salt concentrations. Compared with the emulsions, the emulsion gels exhibit some greatly improved characteristics, including improved oxidative stability of lipids and controlled release for bioactives, and thus may have more potential when applied as the carriers for bioactives. Taking this into consideration, the heat-set emulsion gels are clearly not suitable as the carrier for heat-labile bioactives, while those obtained by cold-set techniques without heat treatment will be much more favorable, especially for food formulations containing heat-sensitive ingredients. Manoi and Rizvi (2009) used a novel supercritical fluid extrusion (SCFX) process to texturize whey protein concentrate (WPC) into a product with cold-setting gel characteristics that was stable over a wide range of temperature. The emulsifying activity and emulsion stability indices of texturized WPC (tWPC) and its ability to prevent coalescence of O/W emulsions were evaluated and compared with the commercial WPC80. The cold, gel-like emulsions were prepared at different oil fractions ($\varphi=0.20$–0.80) by mixing oil with the 20 wt.% tWPC dispersion at 25°C and evaluated using a range of rheological techniques. The microscopic structure of cold, gel-like emulsions was also observed by CLSM. The results revealed that the tWPC showed excellent emulsifying properties compared to the commercial WPC in slowing down emulsion-breaking mechanisms such as creaming and coalescence. Very stable with finely dispersed fat droplets and homogeneous O/W gel-like emulsions could be produced. Liu and Tang (2011) also reported another novel process to produce cold, gel-like emulsions at various φ values in the range 0.2–0.6 from heat-treated whey protein dispersions by microfluidization emulsification. The emulsifying properties of the emulsions formed at low-protein concentrations less than 1% (w/v) were evaluated. The rheological and microstructural characteristics of the gel-like emulsions were characterized using steady and dynamic rheological measurements and a CLSM. CLSM analyses confirmed close relationships between rheological properties and gel network structures. The formation of the gel-like network structure was closely related to the high emulsifying efficiency by microfluidization. This kind of novel gel-like emulsion might exhibit great potential and be applicable in food formulations, for example, as a kind of carrier for heat-labile and active ingredients.

3.7 Nuclear Magnetic Resonance (NMR) Techniques

There are different types of measurements to characterize emulsion physicochemical properties that involve NMR measurements: evaluation of relaxation time T_2 and determination of particle size distribution.

Time-domain low-resolution NMR (TD-NMR) relaxation experiments allow the distribution of the T_2 populations in the emulsions to be determined. ^1H spin–spin relaxation experiments are usually performed using the Carr-Purcell-Meiboom-Gill (CPMG) pulse sequence. Typically, a 90° pulse for 2.8 μs followed by a 180° pulse for 10.0 μs with a recycle delay of 2 s is applied to samples. A true T_2, devoid of any other contribution except that from the sample spin energy exchange, is the time constant of the exponential that describes the envelope of the echo decay in a CPMG experiment. A dynamically heterogeneous sample can have its echo decay analyzed according to a multiple exponential function:

$$M = M_0 \left[K_1 e^{-\tau/T_{21}} + K_2 e^{-\tau/T_{22}} + K_3 e^{-\tau/T_{23}} + .. + K_n e^{-\tau/T_{2n}} \right]$$

from which different T_{2i} values, corresponding to dynamically different populations within the sample, can be calculated. The signal decay represents an envelope of several T_{2i}, from populations of different motilities that can be described by their respective T_{2i} values. To do this, data are fitted initially with the sum of discrete exponential functions, with the best fit given by both the visual adherence of the fitted function to the experimental curve and the presence of a less structured, more random residual curve. With this approach, Silva et al. (2010) were able to identify at least three contributions of different T_{2i} for a rumen or soy protein emulsions corresponding to three major distinct populations of protons in respect to their mobility. The aim of their work was to study the effects of extrusion cooking on rumen's and soy proteins' behavior as emulsion stabilizers. The results showed that extrusion was able to upgrade functional properties of rumen protein and improve some characteristics of the emulsions formed with it. The improved emulsion behavior promoted by rumen extrusion was explained by the molecular reorganization of protein after processing that considerably increased its hydrophobic surface, increasing its emulsion capacity and improving the emulsion structure. According to the authors, extrusion can promote the use of rumen, a byproduct waste from the meat industry, in human nutrition by partially replacing soy protein in food emulsions.

In food science, the description of droplet sizes in quantitative terms is of importance for quality control and process monitoring as well as fundamental food research. The size of oil droplets in oil–water (O/W) emulsions such as mayonnaise and dressings may affect the off-flavor development, flavor release, structure, and appearance of the products. Particle sizing instruments based on NMR utilize interactions between radio waves and the nuclei of hydrogen atoms to obtain information about the microstructure of emulsions (McClements 2007). An emulsion is placed in a static magnetic field gradient and a series of radio frequency pulses is applied to it. These pulses cause some of the hydrogen nuclei in the sample to be excited to higher energy levels, which leads to the generation of a detectable NMR signal. The amplitude of this signal depends on the movement of the nuclei in the sample: The farther the nuclei move during the experiment, the greater the reduction in the amplitude. A measurement of the reduction in signal amplitude with time can therefore be used to study molecular motion. In a bulk liquid, the distance that a molecule can move in a certain time is governed by its translational diffusion coefficient. When a liquid is contained within an emulsion droplet, its diffusion is restricted because of the presence of the interfacial boundary. If the movement of a molecule in a droplet is observed over relatively short times, the diffusion is unrestricted, but if it is observed over longer times, the diffusion is restricted because the molecule cannot move farther than the diameter of the droplet. By measuring the attenuation of the NMR signal at different times, it is possible to identify when the diffusion becomes restricted and thus estimate the droplet size distribution using a suitable mathematical model (McClements 2007). Commercial instruments based on NMR are sensitive to particle sizes between about 200 nm and 100 μm. These instruments can analyze emulsions with particle concentrations ranging from around 1 to 80 wt.% and can therefore be used to analyze many food emulsions without the need for sample dilution. The NMR technique is particularly useful at determining the actual size of the individual droplets in flocculated emulsions (rather than the floc size), as this technique relies on the molecular movement of water molecules within droplets and it detects size increases in the droplets themselves and not the clustering of droplets, thereby differentiating between coalescence and flocculation/coagulation. NMR-restricted diffusion measurements carried out on the continuous phase of emulsions can also be used to provide information about the structural organization of the droplets within flocs. Like ultrasonic spectrometry, NMR is non-destructive and can be used to analyze emulsions that are concentrated and optically opaque (McClements 2007). The absence of any sample pretreatment other than temperature equilibration makes the technique extremely useful to follow changes in the droplet size of various protein-stabilized O/W model emulsions during prolonged storage and temperature cycling. Static light scattering requires extensive sample preparation to break the covalent bonds that form during storage and is therefore not a very suitable alternative for these unstable systems. Furthermore, the interpretation of static light scattering data depends considerably on the correct choice for the value of the complex refractive index, which may be difficult to obtain independently for these complex model emulsions. The droplet size results obtained may also be supplemented with scanning electron microscopy (SEM) imaging (Kiokias et al. 2004b).

The study of partial coalescence in complex emulsions requires a well-considered choice of the model system. A typical paper in the field of emulsion science involves a liquid dispersed phase (e.g., mineral or triacylglycerol oil), which generally forms a stable emulsion. Moreover, in the case of protein-stabilized emulsions, preferably neutral and native systems are studied, which tend to be less prone to partial coalescence, too. In contrast, heated-acidified emulsions prepared with partly crystalline fat, which can serve as model systems for certain cream cheese–type products, are far less stable. Kiokias et al. (2004a) studied the stability of heat-treated and/or -acidified, partly crystalline fat–based, whey protein–stabilized O/W emulsions against partial coalescence during chilled storage (at 5°C) and repeated temperature cycling (three times between 5°C and 25°C). They prepared model O/W emulsions from mixtures of 30% lipid phase consisting of either partly crystalline vegetable fat (1:1 mixture of fully hardened coconut oil and fractionated palm oil, 2.4% solid fat content at 25°C, 14% at 20°C, 73% at 5°C) or liquid vegetable oil (sunflower oil), 4% powder with 75% protein. Arround 10% of the protein is denatured. Experiments focused on the evolution of firmness and droplet size (using pulsed-field gradient NMR and scanning electron microscopy). Besides the effects of denaturation and/or acidification, the influence of the droplet size of the dispersed phase on emulsion stability was investigated also. They formulated emulsions in which the protein content was so high that beforehand destabilization would have not been expected to occur. Nevertheless, that study showed that such emulsions could be destabilized through a reduction in protein functionality by heating and/or acidification. They studied partial coalescence in two ways: in terms of its microscopic and its macroscopic effects. Droplet size was chosen as the microscopic property under investigation. Fat droplet size was known to be related to (partial) coalescence and to textural characteristics of an emulsion gel, and its evolution can be taken as a measure for the long-term stability of the emulsion during chilled storage. Droplet size was studied by means of pulsed-field gradient NMR (pfg-NMR). It was found that heat treatment or acidification before emulsification led to unstable emulsions during temperature cycling, whereas heat treatment after acidification resulted in stable emulsions.

Many water-in-oil (W/O) emulsion-based processed foods (e.g., whipped toppings, table spreads, and sauces) are frozen to improve their long-term storage before thawing for further processing or consumption. In the food industry, freeze/thaw-induced destabilization is considered unacceptable, as it leads to compromised consumer acceptability. Both interfacially active and continuous-phase crystals may play a role in W/O emulsion stability. With the former, colloids collect at the emulsion droplet interface, "anchor" themselves onto the droplet surface, and provide a physical barrier to coalescence. In foods, so-called Pickering species may originate by way of surfactant solidification at the interface [e.g., monoacylglycerols (MAGs)] and/or the migration of previously formed crystals toward the droplet interface. Crystals lacking any surface activity, but that are present at a sufficiently high concentration, will form a plastic network throughout the continuous phase of the emulsion, thus encasing the dispersed phase and reducing droplet diffusion and sedimentation. Oil-in-water (O/W) emulsions are prone to freeze/thaw destabilization, given the large volume expansion of the continuous aqueous phase during freezing. When the oil droplets crystallize prior to the continuous aqueous phase, the emulsion will destabilize due to the formation of a network of aggregated crystalline droplets that coalesce as the dispersed fat phase melts. Ghosh and Rousseau (2009) investigated the relationship between crystallization and the stability of continuous- and dispersed-phase W/O emulsions. Two continuous oil phases were selected based on the order of their crystallization events, either before (coconut oil) or after (canola oil) the crystallization of the emulsified water phase. As well, two surfactants were chosen based on their phase behavior: molten [polyglycerol polyricinoleate (PGPR)] or solid [glycerol monostearate (GMS)]. In so doing, continuous-phase crystallization was isolated from that of the dispersed phase as well as that of the surfactant phase, leading to a clear delineation of the role of interfacial crystallization versus network crystallization and the role of aqueous phase crystallization on stability. Their hypothesis was that Pickering crystals formed by GMS would better stabilize the dispersed aqueous phase than molten PGPR, irrespective of whether the continuous phase was crystallized or not. Emulsion stability was assessed with pulsed-field gradient NMR droplet size analysis, sedimentation, microscopy, and differential scanning calorimetry. This study demonstrated that the sequence of crystallization events (i.e., dispersed after continuous phase or vice versa) and the physical state of the surfactant at the oil–water interface strongly impact the freeze–thaw stability of water-in-oil emulsions. With the materials and experimental conditions studied here, freeze–thaw destabilization of W/O emulsions was most apparent with a liquid-state emulsifier and a continuous oil phase that solidified prior to the dispersed phase (the PGPR-CNO emulsion). Conversely, emulsions stable to freeze–thaw cycling were obtained with emulsifier crystallized at the oil–water interface (the GMSCNO emulsion) or in emulsions where the continuous phase crystallized after the dispersed aqueous phase (the PGPR-CO and GMSCO emulsions). These results showed that the relevance of the choice of surfactant and composition of the continuous oil phase strongly impact the stability of W/O emulsions to freeze–thaw cycling.

3.8 Optical Methods

3.8.1 Dynamic Light Scattering (DLS)

Light scattering (LS) techniques have been applied to many areas of research. There are several reasons for the rapid development of LS methodology. It is noninvasive; the sample to be studied does not need to be prepared in any way (except by dilution), so there are few experimental artifacts. It is relatively easy to use and gives immediate results; measurements may be made in a few seconds if necessary; and depending on the sophistication of the analysis, information on the size distribution is obtained within a short calculation time. The theory of light scattering for a few well-defined models, such as hard spheres and rods, is well established, and there are computer models available to convert experimentally obtained data into size distributions. It is relatively inexpensive to set up and to run. Additionally, results are very repetitive. For all these reasons, dynamic light scattering (DLS) techniques have been used to study the stability of suspensions or emulsions and to estimate the dispersed-phase properties such as particle size and volume fraction. However, these methods require highly transparent, and in general highly diluted, samples. Concentrated suspensions, such as milks or food emulsions, may change if they are diluted into inappropriate media, and great care must be taken to ensure that the particles under study retain the structures they have in the original foods. In addition, the high degree of dilution required generally makes it impossible to study processes that occur only in concentrated media, such as gelation or phase separation. For this reason, DLS techniques have to be associated with other techniques to study emulsion stability in food research (Alexander and Dalgleish 2006). In LS studies, the intensity of the light arriving at the detector at any instant depends on the interference pattern created by the scattered light from all of the particles in the scattering volume. This intensity fluctuation provides information on the sizes and size distributions of the suspended particles, or information on the viscosity of the sample if the particle sizes are known. Usually, this technique measures some property of particles in the analyzed system and assumes that this refers to a sphere, hence deriving one unique number (the diameter of this sphere) to describe the particle. This ensures that only one number can be used to describe a 3D particle and make it unnecessary to describe the particle with three or more numbers, which, although more accurate, is inconvenient for some purposes, such as, for example, the management of products in the food industry. The distribution may be expressed in different ways depending on the application: the number of particles with a specific diameter; the surface mean diameter, also called the Sauter mean diameter ($d_{3,2}$); or the volume mean diameter, also called the De Brouckere mean diameter ($d_{4,3}$). In a catalytic reaction, for example, a comparison of particles on the basis of surface area would be very appropriate because it would be expected that the higher the surface area, the higher the activity of the catalyst. In this case, results would be expressed in $d_{3,2}$ values. According to Relkin and Sourdet (2005), the $d_{4,3}$ parameter is more sensitive to fat droplet aggregation than $d_{3,2}$ (surface weighted diameter). Thus, for evaluation of the structural stability of a food emulsion against flocculation/coalescence, $d_{4,3}$ would be the selected parameter. The advantage of using the parameters $d_{3,2}$ and $d_{4,3}$ may be understood by knowing that in both cases, calculations of the means and distributions do not require knowledge of the number of particles involved. Particle counting is normally only carried out when the numbers are very low (in the ppm or ppb regions) in applications such as contamination, control, and cleanliness. When the distribution of a system with many small particles and a few big particles is expressed in number a monomodal graph is obtained. On the contrary if for the same system the distribution is expressed in volume two peaks, that is, a bimodal distribution is obtained. Big particles are more notorious when the distribution is expressed in volume.

Relkin and Sourdet (2005) studied four emulsions, stabilized by milk proteins, before and after application of whipping and freezing procedures. The emulsions were different in the whey protein/casein ratio and degree of heat denaturation before mixing with the other ingredients. They found that emulsions with a high surface protein concentration and a low crystalline fat content showed greater droplet aggregation under whipping and freezing. Their results indicated that differences in sensitivity to fat droplet aggregation were not caused by differences in resistance to protein displacement from the droplet surface in thawed whipped emulsions, but by low crystalline fat content prior to whipping and freezing. Data collected in this study were discussed in terms of the effects of partial replacement of "native" whey proteins by casein or heat-denatured whey proteins on partial coalescence of fat droplets in whipped-frozen emulsions.

For the preparation of milk whipped creams or ice creams, the lipid ingredient comes mostly from milk fat (in the United States) or vegetable fats (palm oil or palm kernel oil) in other countries. These lipids consist of a wide diversity of fatty acids and triacylglycerols (TAG), each characterized by its own melting temperature.

Their chemical and physical properties may be modified by fractionation or hydrogenation, leading to different characteristic melting profiles and crystallization kinetics (Herrera et al. 1999). Bazmi and Relkin (2006) studied the thermal transitions and physical stability of oil-in-water emulsions containing different milk fat compositions, arising from anhydrous milk fat alone (AMF) or in mixture (2:1 mass ratio) with a high melting temperature fraction (HMF) or a low melting temperature fraction (LMF). The droplet size distribution of these ice cream model systems was evaluated by dynamic light scattering measurements and fluorescence microscopic observations. The results indicated differences in fat droplet aggregation-coalescence under freeze–thaw procedures, depending on the emulsion's fat composition. It appeared that under quiescent freezing, emulsions containing AMF–LMF were much less resistant to fat droplet aggregation-coalescence than emulsions containing AMF or AMF–HMF. Their results suggested that the fat droplet liquid–solid content was very relevant to the emulsion's stability.

Semenova et al. (2009) investigated the relationship between the ratio of sodium caseinate/dextran sulfate used and emulsion stability by combining static and dynamic light scattering. Various structural and thermodynamic parameters have been determined for the complex particles formed from sodium caseinate (0.5 wt/vol.%)/dextran sulfate (0.01, 0.1, or 1.0 wt/vol.%) in aqueous solution at pH = 6.0. The polysaccharide contents refer, respectively, to three polysaccharide/protein molar ratios ($R = 1$, 10, and 100) calculated on the basis of the measured values of the weight-averaged molar masses of sodium caseinate particles and dextran sulfate molecules. The complexes were prepared by mixing together the two biopolymer components in bulk solution or bringing them together at the interface in a protein-stabilized foam. The results indicate dissociation of the original sodium caseinate particles in response to associative interactions with excess amounts of negatively charged polysaccharide. A significant difference was observed between properties of complexes formed in solution and those formed at the interface, especially for $R = 100$. Semenova and colleagues found a possible correlation between the structures of these complexes and the stability properties of oil-in-water emulsions containing the same biopolymers, which was described elsewhere (Jourdain et al. 2008).

Zheng et al. (2011) reported a novel fish oil O/W microemulsion (formed by nanodroplets) system formulated with food acceptable components, Tween 80, ethyl oleate, fish oil, and water. The fish oil used was rich in docosahexaenoic acid (DHA) and eicosapentaenoic acid (EPA), which are known to play a significant role in nervous system activity, cognitive development, memory-related learning, neuroplasticity of nerve membranes, synaptogenesis, and synaptic transmission. The systems were investigated using dynamic light scattering (DLS), transmission electron microscopy, and rheological methods. The obtained results indicated that the particle sizes of spherical droplets in microemulsions depend significantly on the total oil-phase content, varying from 5 to 198 nm. The rheological measurements showed that all studied microemulsions followed shear thinning behavior. Release experiments confirmed that the microemulsion system is potentially useful as a release/delivery system of fish oil.

3.8.2 *Diffusing Wave Spectroscopy (DWS)*

In recent years, great strides have been made in our understanding of colloidal interactions in food emulsions. Traditional experimental techniques such as light scattering, rheology, microscopy, and turbidity have been crucial in determining the forces at play in complex food structures. In the past 20 years, a new light scattering technique called *diffusing wave spectroscopy (DWS)* has been increasingly employed in the study of highly turbid food media. Its main advantage is the ability to extract structural and dynamic information between food components in a noninvasive way. DWS is now gaining pace in its application in the food industry. Much of this success can be attributed to the relatively simple theoretical model and rather inexpensive and straightforward experimental setup. DWS is very similar to traditional dynamic light scattering (DLS) since both techniques follow the temporal fluctuations of intensity of a speckle of scattered light. In both cases, this fluctuation reflects the dynamics of the scattering particles. This fluctuation has a characteristic time scale inversely proportional to the particle diffusion coefficient; for this reason, it can provide information about colloidal particles in the range of tens of nanometers to a few microns. However, conventional light scattering requires the sample to be highly diluted (to be in what is called the *single scattering* regime), and this can severely restrict its use in industrially relevant systems. In contrast to DLS, DWS must operate in a highly turbid medium (or *multiple scattering* regime) because it treats the photon path through the sample as a diffusive process. Similar to DLS, DWS is able to provide information about the local dynamics of particle dispersions without restrictions on particle concentration and turbidity. This is highly desirable in systems such as foods and soft materials, where most of the colloidal dispersions (milk, salad dressing, acidified dairy drinks, etc.) are usually quite opaque. There are two different applications of DWS: the backscattering mode and the forward-scattering mode. The main difference between them arises in the placement of the collector system (lens and fiber optics). In the first case, the collector and detector are placed on the same side as the incident laser light. In the latter case, the collector and detector are placed on the opposite side of the incident light; that is, the scattered light must travel the whole thickness of the sample, L, before it is detected and analyzed. There is a formal similarity between the two methods, but there are significant differences in the interpretation of the results. This topic was extensively reviewed by Corredig and Alexander (2008). The interest in using the method is in studying changes in the behavior of a suspension when it is destabilized by one means or another (Alexander and Dalgleish 2006; Liu et al. 2007b).

Structure is an important parameter for many food products. To control the final structure of a food product, one has to understand not only the structuring properties of different food components but also the influence of the different processing steps. Gelation has been described as the growth of clusters of particles leading to a space-filling network. This process succeeds only if the particles are present above a critical concentration, the attractive interparticle interactions are of sufficient strength, and the applied stress is sufficiently low. In general, the dynamics of colloidal systems that are driven away from equilibrium by the application of shear flow are determined by the interplay of the intrinsic relaxation of the system with shear flow. The particle motion is often studied using light scattering techniques such as dynamic light scattering. An extension of this technique for concentrated, turbid samples is diffusing wave spectroscopy (DWS). DWS has been successfully applied to the study of interactions of dairy proteins as well as the kinetics of aggregation in gelling systems under different destabilizing conditions. Ruis et al. (2008) studied the influence of shear flow on the acid-induced aggregation of a food-based system that consists of an oil-in-water (O/W) emulsion stabilized by sodium caseinate. Slow acidification down to pH 4.6, equal to the isoelectric point of sodium caseinate, causes the aggregation of emulsion droplets. Understanding the aggregation behavior and the effect of shear as a function of the pH of such emulsions is relevant to the food industry, especially the dairy industry. Their results showed that the emulsion droplets in the food-related emulsion were uniformly dispersed at neutral pH. Upon acidification down to a pH of 5.2, the emulsion showed Newtonian behavior with constant viscosity over the whole pH range. At pH 5.17, independent of the applied shear rate during acidification, the viscosity suddenly increased. From this point on, the emulsion showed shear-thinning behavior. In the DWS method, an important parameter is the photon-transport mean free path, l^*. The value of l^* can, in the absence of absorption, be determined using the average intensity of the transmitted light of a sample combined with the transmission, T, of a reference sample with known l^* by

$$T = \frac{I}{I_0} = \frac{\left(5l^*/3L\right)}{\left(1+4l^*/3L\right)}$$

In this equation, the transmission is equal to the ratio of the incoming and transmitted energy flux, which is related to the laser (I_0) and the transmitted (I) average light intensities, and L is the path length. For completely noninteracting scatterers, which are spatially completely uncorrelated, the value of l^* depends on the wavelength of the laser light and on the size, concentration, and optical contrast of the scatterers with the dispersion medium. The photon-transport mean free path (l^*) was not influenced by the applied shear rate and did not change down to pH 5.2. Close to this pH, l^* increased, and at a lower pH (5.05), l^* started to fluctuate. Assuming that the convective motion and the Brownian motion are independent of each other, the mean-square displacement as a result of Brownian motion was determined. From this, the sol–gel point and the radius of the aggregates at this point as a function of the shear rate were determined. The results indicated that the radius of the aggregates at the sol–gel transition decreased with increasing shear rate and suggested that shear will result in a more open structure of the network formed by the aggregates. This study shows that DWS is a promising technique for the study of food systems, even under shear, that are normally very hard to probe by light scattering techniques because of their turbidity.

Polysaccharides are often added to food systems as they impart unique and desirable textures and micro-structural properties. Pectin, in particular, is commonly used as a food ingredient because of its unique properties such as water-holding capacity, viscosity improvement, and gelling properties (Bonnet et al. 2005; Liu et al. 2007c). Pectin is often added to milk as it provides stability to acidified milk products against whey separation. Pectin is an anionic polysaccharide found in plant cell walls and is most widely extracted from citrus peel and apple pomace. It consists of water-soluble, relatively elongated polymeric molecules with carboxyl groups. Its structure is formed by rhamnogalacturonans as well as homopolymeric poly-α-(1$\rightarrow$4)-D-galacturonic acid with α-(1$\rightarrow$2)-L-rhamnosyl-(1$\rightarrow$4)-D-galacturonosyl sections containing branch points with mostly neutral side chains. Some of the carboxyl acid groups in galacturonans can be methyl-esterified and the amount that is substituted is referred to as the *degree of esterification* Pectins with a degree of esterification of 43% or more are classified as high-methoxyl pectins (HMP), whereas those with a lower number of methyl esters are called low-methoxyl pectins (LMP). Acero Lopez et al. (2009) investigated the effect of the addition of different HMP concentrations on the stability of skim milk and on the renneting kinetics of casein micelles at pH 6.7. Noninvasive techniques such as diffusing wave spectroscopy (DWS) and ultrasonic spectroscopy (US) were used to closely observe the structure formation during renneting in the presence of HMP. These two techniques allow for in situ measurements of sol–gel transitions without the need for dilution and provide information about particle size and mobility, as well as interparticle interactions that might occur under destabilizing conditions. Their results showed that at low-HMP concentrations, the casein micelles' aggregation behavior was similar to that of skim milk, although changes could be noted in the microstructure of the renneted gels. At HMP concentrations between 0.1% and 0.15%, phase separation kinetics were slower than the rennet-induced aggregation, and different microstructures formed caused by different dynamics of interactions between casein micelles present in HMP-depleted flocs. Higher amounts of HMP failed to create a continuous gel, as phase separation occurred at a faster rate than rennet aggregation. Acero Lopez et al.'s (2009) work highlights the importance of noninvasive techniques in the study of concentration-dependent phase-separating and -aggregating systems, as only with observations in situ is it possible to determine new ways to control the structuring of protein–polysaccharide mixed systems. These authors also proved that DWS and the attenuation of sound measured by US were in good agreement, and both identified well the structural changes occurring in the emulsions under destabilizing conditions (Liu et al. 2008; Gaygadzhiev et al. 2009; Acero Lopez et al. 2010; Chappellaz et al. 2010).

3.8.3 Turbiscan

Emulsions have been studied by numerous techniques, such as particle sizing, microscopy, and rheology, among others, to characterize their physical properties. Most of these techniques involve some form of dilution. This dilution disrupts some

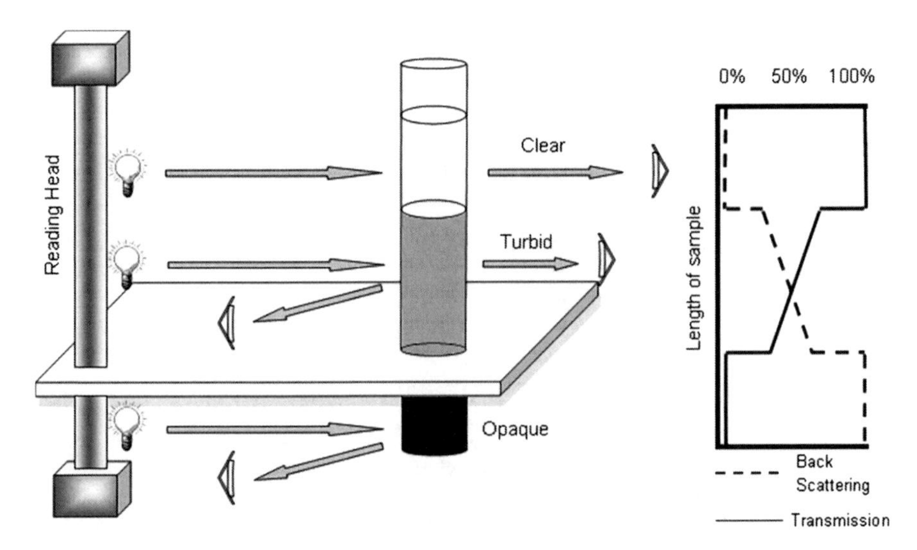

Fig. 3.2 Scheme of the TMA 2000 analyzer

structures such as flocs that contribute to destabilization. The ability to study the stability of food emulsions in their undiluted forms may reveal subtle nuances about their stability. A relatively recently developed technique, the Turbiscan method, allows us to scan the turbidity profile of an emulsion along the height of a glass tube filled with the emulsion, following the fate of the turbidity profile over time. These profiles constitute the macroscopic fingerprint of the emulsion sample at a given time. The analysis of the turbidity profiles leads to quantitative data on the stability of the studied emulsions and allows us to make objective comparisons among different emulsions (Chauvierre et al. 2004).

Figure 3.2 shows a scheme of the vertical scan analyzer Turbiscan MA 2000. This equipment allows the optical characterization of any type of dispersion (Pan et al. 2002; Mengual et al. 1999). The reading head is composed of a pulsed near-IR light source ($\lambda = 850$ nm) and two synchronous detectors. The transmission detector receives the light, which goes through the sample (0°), while the backscattering detector receives the light back-scattered by the sample (135°). The Turbiscan head acquired T and BS data every 40 μm along the vertical length of the cell. Therefore, the scan of a 60-mm height sample provides patterns, including 1,500 points of measurement, in less than 20 s. Thus, by repeating the scan of a sample at different time (t) intervals, the stability or instability of dispersions can be study in detail. The samples are placed in a flat-bottomed cylindrical glass measurement cell and scanned from the bottom to the top. The backscattering (BS) and transmission (T) profiles as a function of the sample height (total height = 60 mm) are obtained in quiescent conditions usually at ambient temperature. The profiles allow the calculation of creaming, sedimentation, or phase separation rates, as well as flocculation, and the mechanism making the dispersion unstable can be deduced from the transmission or backscattering data (Chauvierre et al. 2004).

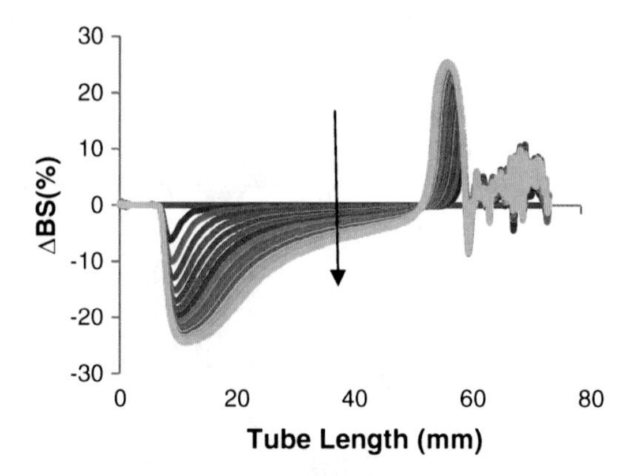

Fig. 3.3 Changes in back scattering (BS) expressed in reference mode profiles (subtracting the profile at $t=0$), as a function of the tube length with storage time (the emulsion was stored for 1 week; *arrow* denotes time) in quiescent conditions. It was formulated with 10 wt.% sunflower oil as fat phase, 20 wt.% sucrose added to the aqueous phase, and a concentration of sodium caseinate (NaCas) of 0.5 wt.%. Tube length = 65 mm

The curves obtained by subtracting the BS profile at $t=0$ from the profile at t ($\Delta BS = BS_t - BS_0$) display a typical shape that allows a better quantification of creaming, flocculation, and other destabilization processes. Figure 3.3 gives an example of a typical profile of an emulsion that destabilized mainly by creaming of small particles in direct and reference mode. Creaming was detected using the Turbiscan as it induced a variation of the concentration between the top and bottom of the cell. The droplets moved upward because they had a lower density than the surrounding liquid. When creaming takes place in an emulsion, the ΔBS curves show a peak at heights between 0 and 20 mm. The variation of the peak width, at a fixed height, during the studied time, can be related to the kinetics of migration of small particles (Mengual et al. 1999). The creaming destabilization kinetics may be evaluated by measuring the peak thickness at 50% of the height at different times (bottom zone). The slope of the linear part of a plot of peak thickness vs. t gives an indication of the migration rate. Figure 3.4 gives an example of peak thickness vs. t for the emulsion in Fig. 3.3.

For emulsions that destabilize mainly by flocculation, BS mean values (BS_{av}) change with the increase in particle size. Flocculation is followed by measuring the BS_{av} as a function of storage time in the middle zone of the tube. As was theoretically demonstrated by Mengual et al. (1999), the BS intensity decreased as the particle size increased [when particle size is higher than the wavelength (λ) of the incident light]. It should be mentioned that if the particle size is lower than λ of the incident light, BS increases with particle size. This phenomenon was used by several authors to determine flocculation kinetics (Chauvierre et al. 2004; Palazolo et al. 2005). The optimum zone is the one not affected by creaming (bottom and top of the tube), that is, the 20–50-mm zone. Figure 3.5 shows a typical profile of an emulsion that destabilized mainly by flocculation in direct and reference mode. The actual profile of a flocculated emulsion shows a significant change in BS_{av} in the middle of the tube and its shape at the top of the tube is different from that of migration of individual particles, which is a consequence of the migration of flocs formed

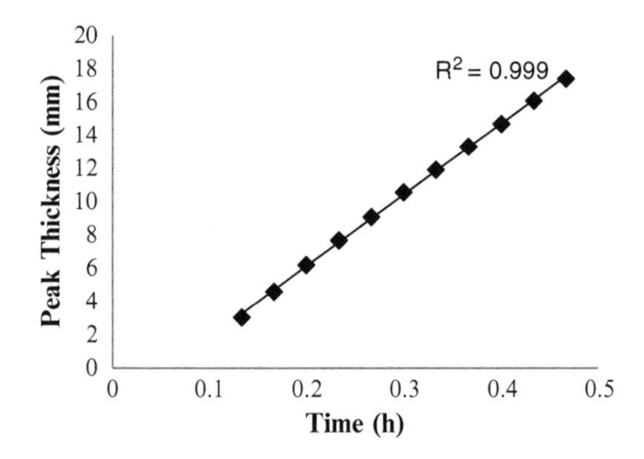

Fig. 3.4 Peak thickness measured at 50% of the height at different times in the bottom zone of the tube, for emulsion in Fig. 3.3 stored in quiescent conditions at 22.5°C, monitored over 30 min. Tube length = 65 mm

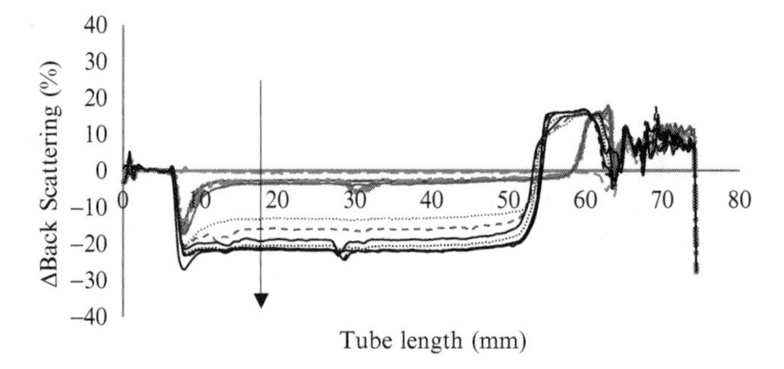

Fig. 3.5 Changes in back scattering (BS) in reference mode profiles, as a function of the tube length with storage time (samples were stored for 1 week; *arrow* denotes time) in quiescent conditions for the emulsion formulated with 10 wt.% sunflower oil as the fat phase, no sugar added to the aqueous phase, and a concentration of sodium caseinate (NaCas) of 5.0 wt.%. Tube length = 65 mm

during flocculation. In the case of flocculation, the BS profile in reference mode does not show a peak at the bottom of the tube (0–20 mm), which indicates that there is no detectable migration of small particles.

Several food systems have been studied using a Turbiscan. Álvarez Cerimedo et al. (2010) described the stability behavior of emulsions formulated with high concentrations of sodium caseinate and trehalose. The emulsion were formulated with 10 wt.% oil (concentrated fish oil, CFO; sunflower oil, SFO; or olive oil, OO), sodium caseinate concentrations varying from 0.5 to 5 wt.%, giving oil-to-protein ratios of 20–2, and 0, 20, 30, or 40 wt.% aqueous trehalose solution. Those systems showed a more complicated behavior than that reported in literature. The oil-to-protein ratio that gave stability changed with processing conditions and formulation of the aqueous phase. Many studies in the literature evaluated emulsions by visual observation. By using this approach, all our emulsions would be considered stable since no phase separation occurred after a week in quiescent conditions at 22.5°C. This is in agreement with the fact that the transmission detector received no light during the time emulsions were analyzed. However, the BS detector was able to quantify creaming or flocculation in these systems. Concentrations below 0.5 wt.% NaCas seemed to be below the ones required for saturation monolayer coverage since the creaming rate is greater for 0.5 wt.% than for 1 wt.% NaCas. Systems stabilized by these oil-to-protein ratios most likely destabilized by time-dependent bridging flocculation. Further addition of protein led to high instability. The 2–4 wt.% NaCas range was the worst situation. At these protein contents, the creaming stability of the system was greatly reduced probably due to depletion flocculation of protein-coated droplets by unabsorbed submicellar caseinate. Although it was reported that at protein contents around that required for saturation monolayer coverage, the system was very stable toward creaming and coalescence (Dickinson 1999), there was no such oil-to-protein ratio in our systems between bridging flocculation and depletion flocculation destabilization processes. At even higher protein contents (5 wt.% NaCas concentration), emulsions were very stable especially when the aqueous phase contained trehalose. In these conditions, the systems remained in the liquid state for at least a week, fully turbid; that is, no changes were noticed by visual analysis, and no creaming or flocculation was detected by the Turbiscan. Although there was no gel formation when emulsions were kept at 22.5°C for a week, they were very stable. Confocal images obtained after 24 h at 22.5°C were in agreement with Turbiscan data. Turbiscan allowed quantification of creaming and coalescence kinetics and more objective comparisons.

Cabezas et al. (2011) evaluated the emulsifying properties of different modified sunflower lecithins in oil-in-water (O/W) emulsions. They studied five modified sunflower lecithins, which were obtained by deoiling (deoiled lecithin), fractionation with absolute ethanol (PC- and PI-enriched fractions), and enzymatic hydrolysis with phospholipase A2 from pancreatic porcine and microbial sources (hydrolyzed lecithins). Modified lecithins were applied as an emulsifying agent in O/W emulsions (30:70 wt/wt), ranging from 0.1% to 2.0% (wt/wt), and the stability of different emulsions was evaluated through the evolution of backscattering profiles (%BS), particle size distribution, and mean particle diameters ($D_{4,3}$, $D_{3,2}$). The PC-enriched fraction and both hydrolyzed lecithins presented the best emulsifying properties against the main destabilization processes (creaming and coalescence) for the analyzed emulsions. These modified lecithins represent a good alternative for the production of new bioactive agents.

Juliano et al. (2011) enhanced the creaming of milk fat globules in milk emulsions by the application of ultrasound and detected instability using the Turbiscan method. Coarse and fine recombined emulsions and raw milk were subjected to ultrasound for 5 min at 35°C. Scans were performed over a 10-min period after removing the samples from the ultrasonic setup. For each model system, an unsonicated control was treated exactly in the same way as the sonicated samples. They were filled into the Turbiscan tube, put into the water bath, and after 5 min placed into the Turbiscan; measurements were performed for an additional 10 min. Another way to quantify creaming destabilization by Turbiscan is the one used by Juliano et al. (2011). Due to the increased backscattered light at the layers with a higher number of dispersed particles, it is possible to identify high-cream-concentration regions. Since the Turbiscan can measure highly concentrated solutions (up to

95% v/v), no significant nonlinear effects are introduced due to high particle concentrations. The backscattering is directly proportional to the square root of the concentration only when the particle size is unchanged. To compare the different creaming extents, the ratio of the area of the peak corresponding to the separated cream phase was calculated with respect to the total backscattering measured in the sample. The cream phase was defined as the curve area, starting where the backscattering value is greater than 1.5% in comparison to the baseline value representing the bulk liquid. The 1.5% value represents the measurement error of backscattering determined for the Turbiscan equipment. For comparison between different measurements, the area under the cream phase curve was related to the area under the whole curve. The creaming extent (CE) can be expressed as

$$\mathrm{CE} = \frac{\int_0^L \mathrm{BS}(z)_{\text{cream phase}}\,dz}{\int_0^L \mathrm{BS}(z)\,dz}$$

where z represents the axial position in the tube (origin at the bottom), L the length corresponding to the free surface, and BS(z) the backscattering at position z. This approach assumes that the backscattering, which is related to the number and size of particles, is shifted when a cream phase is formed. When the particle size of the dispersions at the time of measurement is unchanged, the calculated creaming extent provides an indication of the concentration of fat in the creamed phase. However, when the particle size of the dispersions at the time of measurement is increased, then the backscattering may be either an overestimate or an underestimate of concentration. This is because there is a change in the behavior of the backscattering around the wavelength of the incident light, where backscattering increases with particle size up to ~1 μm and then decreases with further increases in particle size. The calculation of a recovery ratio (RR) computed with the following equation allows an assessment of the validity of the calculated creaming extent.

$$\mathrm{RR} = \frac{\int_0^L \mathrm{BS}(z)_{sample}\,dz}{\int_0^L \mathrm{BS}(z)_{control}\,dz}$$

$\mathrm{BS}(z)_{\text{sample}}$ stands for the backscattering of the sample and $\mathrm{BS}(z)_{\text{control}}$ for the corresponding nonultrasonic treated control at the same time. If this value is close to unity, it means that the overall backscattering of the sample and the control match. Assuming no change in particle sizes, a recovery ratio of 1 can be attributed to the fact that all particles may have moved to another position in the tube but were not flocculated or coalesced at the time the measurement was taken. However, if there is a change in particle size, the recovery ratio is the sum of the relative contributions of the differently sized particles to the backscattering. If there is flocculation/coalescence that results in larger particles (i.e., greater 1 μm), backscattering will be reduced, and the recovery ratio can be less than 100%. If the recovery ratio is greater than 100%, this can be because of particle size increase due to flocculation/coalescence under conditions where the particle size of the floccules/coalesced fat is equal to or less than 1 μm. In Juliano et al.'s systems, creaming, as calculated from Turbiscan measurements, was more evident in the coarse recombined emulsion and raw milk compared to that of the recombined fine emulsion. Micrographs confirmed that there were flocculation and coalescence in the creamed layer of emulsion. Coalescence was confirmed by particle size measurement. Particle flocculation and clustering were detected in both the coarse emulsion and raw milk upon sonication, the extent of which was dependent on the conditions of sonication. Turbiscan was a successful tool to follow destabilization processes in milk emulsion systems under high-frequency ultrasonic waves.

There is also a newly delivered Turbiscan online that allows monitoring and characterization of processes involving emulsions, suspensions, or foams, such as emulsification (salad dressings, mayonnaise), crystallization (sugar), or bulking (foams). Buron et al. (2004) reported that the average droplet size calculated from online measurements with Turbiscan were in agreement with light scattering measurements in the same systems. Measurements were accurate in a wide range of particle volume fraction (0–60%) and particle size (0.01–1,000 μm). Pizzino et al. (2009) performed an online light backscattering tracking of the transitional phase inversion of emulsions. In their study, a surfactant/oil/water system (Brij 30/decane/brine) was stirred while continuously changing the temperature in order to induce the swap of the emulsion morphology from O/W to W/O, or vice versa. The transitional phase inversion was detected by monitoring both the electrical conductivity and the light backscattering with Turbiscan (online) equipment. They found that light backscattering was a complementary technique to conductivity measurement to track the emulsion phase inversion. Provided that the experimental conditions are optimized, it may be considered an easy-to-use, noninvasive, and accurate method of detection with some advantages over the classical conductivity measurement: (1) It does not require any addition of electrolyte in the aqueous phase; (2) it provides some information on the drop size; and (3) it works with both O/W and W/O morphologies, whereas conductimetry is blind for the latter case.

Many industrial processes involve shear-induced dispersion or agglomeration of suspensions, especially for the production of pigments, stains, or varnish as well as for the development of cosmetics, pharmaceutical products, or food products. In such processes involving concentrated micron and submicron particle suspensions, the control of flocculation processes and suspension microstructural changes remains a problem of relevance. Now, industrial inspection is usually performed either after dilution or using intrusive techniques that may induce a denaturing of the medium. Bordes et al. (2003) developed a nonintrusive and nondenaturing optical method based on multiple light scattering phenomena. This device was used to describe in situ the physical properties of a flowing concentrated latex aqueous suspension. An experimental setup, comprising the optical sensor, the Turbiscan reading head, directly mounted inside a Couette flow system, was specifically designed to study both shear- and time-dependent microstructural changes to a stabilized latex suspension. Reversible depletion flocculation induced by the addition of a polymer aqueous solution was also investigated. Scattering parameters of the suspension are derived from measurements of the backscattering level within the framework of a mean field scattering model based on the photon diffusion approximation. Shear flow was shown to promote hydrodynamic melting of crystallites in a stabilized concentrated latex suspension. The depletion flocculation kinetic of 24% latex–CMC (sodium carboxymethylcellulose) suspension was further investigated in Couette flow, and the time dependence of the transport mean path was determined during cluster growth. As a conclusion, the rheo-optical system is well suited to study flocculation processes and monitor aggregation kinetics in concentrated suspensions.

The Turbiscan method was also used to study nanosystems. Lemarchand et al. (2003) developed a new generation of polysaccharide-decorated nanoparticles, which has been successfully prepared from a family of PCL-DEX amphiphilic copolymers varying both the molecular weight and the proportion by weight of DEX in the copolymer. According to the authors, they may have potential applications in drug encapsulation and targeting. The nanoparticles were prepared by a technique derived from emulsion–solvent evaporation, during which emulsion stability was investigated using a Turbiscan. This methodology reveals irreversible (coalescence or aggregation) or reversible (creaming or sedimentation) destabilization much earlier than the operator's naked eye. The nanoparticle size distribution, density, ζ-potential, morphology, and suitability for freeze-drying were determined. These studies of emulsion stability yielded much interesting information such as an understanding of the ability of the copolymers to migrate to the solvent–water interface, a means of following the procedure of nanoparticle preparation and of determining the factor of coalescence (F_c) of the emulsion droplets during the evaporation step. This parameter was defined as follows:

$$F_c = \frac{\rho x v}{m} \left(d_{NP} \mid d_d \right)^3$$

where ρ is the density of the PCL-DEX nanoparticles, m the copolymer weight, v the volume of ethyl acetate, d_{NP} the mean diameter of the nanoparticles after solvent evaporation, and d_d the mean diameter of the emulsion droplets. Because of their strongly amphiphilic properties, the PCL-DEX copolymers were able to stabilize O/W emulsions without the need for additional surfactants. Nanoparticles with a controlled mean diameter ranging from 100 to 250 nm were successfully prepared. Zeta-potential measurements confirmed the presence of a DEX coating. Taking into account the actual relevance of nanoscience, this technique has great potential for studying a wide variety of systems.

References

Acero Lopez A, Corredig M, Alexander M (2009) Diffusing wave and ultrasonic spectroscopy of rennet-induced gelation of milk in the presence of high-methoxyl pectin. Food Biophys 4:249–259

Acero Lopez A, Alexander M, Corredig M (2010) Diffusing wave spectroscopy and rheological studies of rennet-induced gelation of skim milk in the presence of pectin and κ-carrageenan. Int Dairy J 20:328–335

Alexander M, Dalgleish DG (2006) Dynamic light scattering techniques and their applications in food science. Food Biophys 1:2–13

Alvarez Cerimedo MS, Huch Iriart C, Candal RJ, Herrera ML (2010) Stability of emulsions formulated with high concentrations of sodium caseinate and trehalose. Food Res Int 43:1482–1493

Bazmi A, Relkin P (2006) Thermal transitions and fat droplet stability in ice cream mix model systems. Effect of milk fat fractions. J Therm Anal Calorim 84:99–104

Beattie JK, Djerdjev A (2000) Rapid electroacoustic method for monitoring dispersion: zeta potential titration of alumina with ammonium poly(methacrylate). J Am Ceram Soc 83:2360–2364

Berg T, Arlt P, Brummer R, Emeis D, Kulicke WM, Wiesner S, Wittern KP (2004) Insights into the structure and dynamics of complex W/O-emulsions by combining NMR, rheology and electron microscopy. Colloids Surf A Physicochem Eng Asp 238:59–69

Berli CLA, Quemada D, Parker A (2002) Modelling the viscosity of depletion flocculated emulsions. Colloids Surf A Physicochem Eng Asp 203:11–20

Binks BP, Desforges A, Duff DG (2007) Synergistic stabilization of emulsions by a mixture of surface-active nanoparticles and surfactant. Langmuir 23:1098–1106

Blonk JCG, Van Aalst H (1993) Confocal scanning light microscopy in food research. Food Res Int 26:297–311

Bonnet C, Corredig M, Alexander M (2005) Stabilization of caseinate-covered oil droplets during acidification with high methoxyl pectin. J Agric Food Chem 53:8600–8606

Bordes C, Snabre P, Frances C, Biscans B (2003) Optical investigation of shear- and time-dependent microstructural changes to stabilized and depletion-flocculated concentrated latex sphere suspensions. Powder Technol 130:331–337

Boyd BJ, Rizwan SB, Dong YD, Hook S, Rades T (2007) Self-assembled geometric liquid-crystalline nanoparticles imaged in three dimensions: hexosomes are not necessarily flat hexagonal prisms. Langmuir 23:12461–12464

Buron H, Mengual O, Meunier G, Cayre I, Snabre P (2004) Optical characterization of concentrated dispersions: applications to laboratory analyses and on-line process monitoring and control. Polym Int 53:1205–1209

Cabezas DM, Madoery R, Diehl BWK, Tomas MC (2011) Emulsifying properties of different modified sunflower lecithins. J Am Oil Chem Soc. doi:10.1007/s11746-011-1915-8

Chanamai R, Coupland JN, McClements DJ (1998) Effect of temperature on the ultrasonic properties of oil-in-water emulsions. Colloids Surf A Physicochem Eng Asp 139:241–250

Chappellaz A, Alexander M, Corredig M (2010) Phase separation behavior of caseins in milk containing flaxseed gum and κ-carrageenan: a light-scattering and ultrasonic spectroscopy study. Food Biophys 5:138–147

Chauvierre C, Labarre D, Couvreur P, Vauthier C (2004) A new approach for the characterization of insoluble amphiphilic copolymers based on their emulsifying properties. Colloids Polym Sci 282:1097–1104

Chen J, Vogel R, Werner S, Heinrich G, Clausse D, Dutschk V (2011) Influence of the particle type on the rheological behavior of Pickering emulsions. Colloids Surf A Physicochem Eng Asp 382:238–245

Chhabra RP, Agarwal S, Chaudhary K (2003) A note on wall effect on the terminal falling velocity of a sphere in quiescent Newtonian media in cylindrical tubes. Powder Technol 19:53–58

Cho YH, McClements DJ (2007) In situ electroacoustic monitoring of polyelectrolyte adsorption onto protein-coated oil droplets. Langmuir 23:3932–3936

Corredig M, Alexander M (2008) Food emulsions studied by DWS: recent advances. Trends Food Sci Technol 19:67–75

Corredig M, Alexander M, Dalgleish DG (2004a) The application of ultrasonic spectroscopy to the study of the gelation of milk components. Food Res Int 37:557–565

Corredig M, Verespej E, Dalgleish DG (2004b) Heat-induced changes in the ultrasonic properties of whey proteins. J Agric Food Chem 52:4465–4471

Dalgleish DG, Spagnuolo PA, Goff DG (2004) A possible structure of the casein micelle based on high-resolution field-emission scanning electron microscopy. Int Dairy J 14:1025–1031

Dalgleish DG, Verespej E, Alexander M, Corredig M (2005) The ultrasonic properties of skim milk related to the release of calcium from casein micelles during acidification. Int Dairy J 15:1105–1112

de Castro SR, Kawazoe Sato AC, Lopes da Cunha R (2012) Emulsions stabilized by heat-treated collagen fibers. Food Hydrocolloid 26:73–81

Dickinson E (1995) Emulsion stabilization by polysaccharides and protein-polysaccharide complexes. In: Stephen AM (ed) Food polysaccharides and their applications. Marcel Dekker, New York, pp 501–515

Dickinson E (1998) Proteins at interfaces and in emulsions: stability, rheology and interactions. J Chem Soc Faraday Trans 94:1657–1669

Dickinson E (1999) Caseins in emulsions: interfacial properties and interactions. Int Dairy J 9: 305–312

Dickinson E (2003) Hydrocolloids at interfaces and the influence on the properties of dispersed systems. Food Hydrocolloids 17:25–39

Dickinson E, Golding M (1997a) Rheology of sodium caseinate stabilized oil-in-water emulsions. J Colloid Interface Sci 191:166–176

Dickinson E, Golding M (1997b) Depletion flocculation of emulsions containing unadsorbed sodium caseinate. Food Hydrocolloids 11:13–18

Dickinson E, Golding M, Povey MJW (1997) Creaming and flocculation of oil-in-water emulsions containing sodium caseinate. J Colloid Interface Sci 185:515–529

Dissanayeke M, Vasiljevic T (2009) Functional properties of whey proteins affected by heat treatment and hydrodynamic high-pressure shearing. J Dairy Sci 92:1387–1397

Dukhin AS, Goetz PJ, Fang X, Somasundaran P (2010) Monitoring nanoparticles in the presence of larger particles in liquids using acoustics and electron microscopy. J Colloid Interface Sci 342:18–25

Dwyer C, Donnelly L, Buckin V (2005) Ultrasonic analysis of rennet-induced pre-gelation and gelation processes in milk. J Dairy Res 72:303–310

Fox PF, Brodkorb A (2008) The casein micelle: historical aspects, current concepts and significance. Int Dairy J 18:677–684

Gancz K, Alexander M, Corredig M (2005) Interactions of high methoxyl pectin with whey proteins at oil/water interfaces at acid pH. J Agric Food Chem 53:2236–2241

Gaygadzhiev Z, Alexander M, Corredig M (2009) Sodium caseinate-stabilized fat globules inhibition of the rennet-induced gelation of casein micelles studied by diffusing wave spectroscopy. Food Hydrocolloids 23:1134–1138

Ghosh S, Rousseau D (2009) Freeze-thaw stability of water-in-oil emulsions. J Colloid Interface Sci 339:91–102

Gulseren İ, Corredig M (2011) Changes in colloidal properties of oil in water emulsions stabilized with sodium caseinate observed by acoustic and electroacoustic spectroscopy. Food Biophys 6:534–542

Gulseren İ, Alexander M, Corredig M (2010) Probing the colloidal properties of skim milk using acoustic and electroacoustic spectroscopy. Effect of concentration, heating and acidification. J Colloid Interface Sci 351:493–500

Heertje I, Vandervlist P, Blonk JCG, Hendrickx H, Brakenhoff GJ (1987) Confocal scanning laser microscopy in food research—some observations. Food Microstruct 6:115–120

Herrera ML, Hartel RW (2001) Unit D 3.2.1–6 lipid crystalline characterization, basic protocole. In: Current protocols in food analytical chemistry (CPFA), vol I. Wiley, New York. ISBN 0-471-32565-1

Herrera M, de Leon GM, Hartel RW (1999) A kinetic analysis of crystallization of a milk fat model system. Food Res Int 32:289–298

Hiemenz PC, Rajagopalan R (1997) Electrostatic and polymer-induced colloid stability. In: Principles of colloid and surface chemistry, 3rd edn. Marcel Dekker, New York, pp 575–624

Jena S, Das H (2006) Modeling of particle size distribution of sonicated coconut milk emulsion: effect of emulsifiers and sonication time. Food Res Int 39:606–611

Jirapeangtong K, Siriwatanayothin S, Chiewchan N (2008) Effects of coconut sugar and stabilizing agents on stability and apparent viscosity of high-fat coconut milk. J Food Eng 87:422–427

Jourdain L, Leser ME, Schmitt C, Michel M, Dickinson E (2008) Stability of emulsions containing sodium caseinate and dextran sulfate: relationship to complexation in solution. Food Hydrocolloids 22:647–659

Juliano P, Kutter A, Cheng LJ, Swiergon P, Mawson R, Augustin MA (2011) Enhanced creaming of milk fat globules in milk emulsions by the application of ultrasound and detection by means of optical methods. Ultrason Sonochem 18:963–973

Kalnin D, Ouattara M, Ollivon M (2004) A new method for the determination of the concentration of free and associated sodium caseinate in model emulsions. Prog Colloid Polym Sci 128:207–211

Karunasawat K, Anprung P (2010) Effect of depolymerized mango pulp as a stabilizer in oil-in-water emulsion containing sodium caseinate. Food Bioprod Process 88:202–208

Keowmaneechai E, McClements DJ (2002) Influence of EDTA and citrate on physicochemical properties of whey protein-stabilized oil-in-water emulsions containing $CaCl_2$. J Agric Food Chem 50:7145–7153

Khalloufi S, Alexander M, Goff HD, Corredig M (2008) Physicochemical properties of whey protein isolate stabilized oil-in-water emulsions when mixed with flaxseed gum at neutral pH. Food Res Int 41:964–972

Kiokias S, Reiffers-Magnani CK, Bot A (2004a) Stability of whey-protein-stabilized oil-in-water emulsions during chilled storage and temperature cycling. J Agric Food Chem 52:3823–3830

Kiokias S, Reszka AA, Bot A (2004b) The use of static light scattering and pulsed-field gradient NMR to measure droplet sizes in heat-treated acidified protein-stabilised oil-in-water emulsion gels. Int Dairy J 14:287–295

Konya M, Dekany I, Eros I (2007) X-ray investigation of the role of the mixed emulsifier in the structure formation in o/w creams. Colloid Polym Sci 285:657–663

Lemarchand C, Couvreur P, Besnard M, Costantini D, Gref R (2003) Novel polyester-polysaccharide nanoparticles. Pharm Res 20:1284–1292

Liu F, Tang CH (2011) Cold, gel-like whey protein emulsions by microfluidisation emulsification: rheological properties and microstructures. Food Chem 127:1641–1647

Liu J, Alexander M, Verespej E, Corredig M (2007a) Real-time determination of structural changes of sodium caseinate-stabilized emulsions containing pectin using high resolution ultrasonic spectroscopy. Food Biophys 2:67–75

Liu J, Corredig M, Alexander M (2007b) A diffusing wave spectroscopy study of the dynamics of interactions between high methoxyl pectin and sodium caseinate emulsions during acidification. Colloids Surf B Biointerfaces 59:164–170

Liu J, Verespej E, Alexander M, Corredig M (2007c) Comparison on the effect of high-methoxyl pectin or soybean soluble polysaccharide on the stability of sodium caseinate-stabilized oil/water emulsions. J Agric Food Chem 55:6270–6278

Liu J, Verespej E, Corredig M, Alexander M (2008) Investigation of interactions between two different polysaccharides with sodium caseinate-stabilized emulsions using complementary spectroscopic techniques: diffusing wave and ultrasonic spectroscopy. Food Hydrocolloids 22:47–55

Macierzanka A, Szeląg H (2006) Microstructural behavior of water-in-oil emulsions stabilized by fatty acid esters of propylene glycol and zinc fatty acid salts. Colloids Surf A Physicochem Eng Asp 281:125–137

Maldonado-Valderrama J, Rodriguez Patino JM (2010) Interfacial rheology of protein-surfactant mixtures. Curr Opin Colloid Interface Sci 15:271–282

Manoi K, Rizvi SSH (2008) Rheological characterizations of texturized whey protein concentrate-based powders produced by reactive supercritical fluid extrusion. Food Res Int 41:786–796

Manoi K, Rizvi SSH (2009) Emulsification mechanisms and characterizations of cold, gel-like emulsions produced from texturized whey protein concentrate. Food Hydrocolloids 23:1837–1847

Marangoni AG, Hartel RW (1998) Visualization and structural analysis of fat crystal networks. Food Technol 52:46–51

Matsumiya K, Nakanishi K, Matsumura Y (2011) Destabilization of protein-based emulsions by diglycerol esters of fatty acids—the importance of chain length similarity between dispersed oil molecules and fatty acid residues of the emulsifier. Food Hydrocolloids 25:773–780

McClements DJ (1999) Emulsion rheology. In: Food emulsions principles, practice and techniques. CRC Press, Washington, DC, pp 235–266

McClements DJ (2007) Critical review of techniques and methodologies for characterization of emulsion stability. Crit Rev Food Sci Nutr 47:611–649

McClements DJ, Coupland JN (1996) Theory of droplet size distribution measurements in emulsions using ultrasonic spectroscopy. Colloids Surf A Physicochem Eng Asp 117:161–170

Medina-Torres L, Calderas F, Gallegos-Infante JA, Gonzalez-Laredo RF, Rocha-Guzman N (2009) Stability of alcoholic emulsions containing different caseinates as a function of temperature and storage time. Colloids Surf A Physicochem Eng Asp 352:38–46

Mengual O, Meunier G, Cayre I, Puech K, Snabre P (1999) Turbiscan MA 2000: multiple light scattering measurement for concentrated emulsion and suspension instability analysis. Talanta 50:445–456

Moschakis T, Murray BS, Biliaderis CG (2010) Modifications in stability and structure of whey protein-coated o/w emulsions by interacting chitosan and gum arabic mixed dispersions. Food Hydrocolloids 24:8–17

Mun S, Cho Y, Decker EA, McClements DJ (2008) Utilization of polysaccharide coatings to improve freeze-thaw and freeze-dry stability of protein-coated lipid droplets. J Food Eng 86:508–518

Murray BS (2011) Rheological properties of protein films. Curr Opin Colloid Interface Sci 16:27–35

Nambam JS, Philip J (2012) Competitive adsorption of polymer and surfactant at a liquid droplet interface and its effect on flocculation of emulsion. J Colloid Interface Sci 366:88–95

Opawale FO, Burgess DJ (1998) Influence of interfacial properties of lipophilic surfactants on water- in-oil emulsion stability. J Colloid Interface Sci 197:142–150

Palazolo GG, Sorgentini DA, Wagner JR (2005) Coalescence and flocculation in o/w emulsions of native and denatured whey soy proteins in comparison with soy protein isolates. Food Hydrocolloids 19:595–604

Pan LG, Tomas MC, Anon MC (2002) Effect of sunflower lecithins on the stability of water-in-oil and oil-in-water emulsions. J Surfact Deterg 5:135–143

Pinfield VJ, Povey MJW, Dickinson E (1995) The application of modified forms of the Urick equation to the interpretation of ultrasound velocity in scattering systems. Ultrasonics 33: 243–251

Pizzino A, Catte M, Van Hecke E, Salager JL, Aubry JM (2009) On-line light backscattering tracking of the transitional phase inversion of emulsions. Colloids Surf A Physicochem Eng Asp 338:148–154

Plucknett KP, Pomfret SJ, Normand V, Ferdinando D, Veerman C, Frith WJ, Norton IT (2001) Dynamic experimentation on the confocal laser scanning microscope: application to soft-solid, composite food materials. J Microsc 201:279–290

Relkin P, Sourdet S (2005) Factors affecting fat droplet aggregation in whipped frozen protein-stabilized emulsions. Food Hydrocolloids 19:503–511

Rosner S, Shalev DE, Shames AI, Ottaviani MF, Aserina A, Garti N (2010) Do food microemulsions and dietary mixed micelles interact? Colloids Surf B Biointerfaces 77:22–30

Ruis HGM, Venema P, van der Linden E (2008) Diffusing wave spectroscopy used to study the influence of shear on aggregation. Langmuir 24:7117–7123

Schokker EP, Dalgleish DG (1998) The shear-induced destabilization of oil-in-water emulsions using caseinate as emulsifier. Colloids Surf A Physicochem Eng Asp 145:61–69

Schokker EP, Dalgleish DG (2000) Orthokinetic flocculation of caseinate-stabilized emulsions: influence of calcium concentration, shear rate, and protein content. J Agric Food Chem 48:198–203

Semenova MG, Belyakova LE, Polikarpov YN, Antipova AS, Dickinson E (2009) Light scattering study of sodium caseinate plus dextran sulfate in aqueous solution: relationship to emulsion stability. Food Hydrocolloids 23:629–639

Silva ACC, Arêas EPG, Silva MA, Arêas JAG (2010) Effects of extrusion on the emulsifying properties of rumen and soy protein. Food Biophys 5:94–102

Stamkulov NS, Mussabekov KB, Aidarova SB, Luckham PF (2009) Stabilisation of emulsions by using a combination of an oil soluble ionic surfactant and water soluble polyelectrolytes. I: emulsion stabilisation and Interfacial tension measurements. Colloids Surf A Physicochem Eng Asp 335:103–106

Stevenson ME, Horne DS, Leaver J (1997) Displacement of native and thiolated b-casein from oil-water interfaces—effect of heating, ageing and oil phase. J Food Hydrocolloids 11:3–6

Sun W, Sun D, Wei Y, Liu S, Zhang S (2007) Oil-in-water emulsions stabilized by hydrophobically modified hydroxyethyl cellulose: adsorption and thickening effect. J Colloid Interface Sci 311:228–236

Tangsuphoom N, Coupland JN (2008) Effect of surface-active stabilizers on the microstructure and stability of coconut milk emulsions. Food Hydrocolloids 22:1233–1242

Tangsuphoom N, Coupland JN (2009a) Effect of surface-active stabilizers on the surface properties of coconut milk emulsions. Food Hydrocolloids 23:1801–1809

Tangsuphoom N, Coupland JN (2009b) Effect of thermal treatments on the properties of coconut milk emulsions prepared with surface-active stabilizers. Food Hydrocolloids 23:1792–1800

Tipvarakarnkoon T, Einhorn-Stoll U, Senge B (2010) Effect of modified Acacia gum (SUPER GUM™) on the stabilization of coconut o/w emulsions. Food Hydrocolloids 24:595–601

Yaghmur A, Glatter O (2009) Characterization and potential applications of nanostructured aqueous dispersions. Adv Colloid Interface Sci 147–148:333–342

Yang R, Gao RC, Cai CF, Xu H, Li F, He HB, Tang X (2010) Preparation of gel-core-solid lipid nanoparticle: a novel way to improve the encapsulation of protein and peptide. Chem Pharm Bull 58:1195–1202

Yang JS, Xie YJ, He W (2011) Research progress on chemical modification of alginate: a review. Carbohydr Polym 84:33–39

Yang JS, Jiang B, He W, Xia YM (2012) Hydrophobically modified alginate for emulsion of oil-in-water. Carbohydr Polym 87:1503–1506

Zheng MY, Liu F, Wang ZW, Baoyindugurong JH (2011) Formation and characterization of self-assembling fish oil microemulsions. Colloid J 73:319–326

Index

M.L. Herrera, *Analytical Techniques for Studying the Physical Properties of Lipid Emulsions*, SpringerBriefs in Food, Health, and Nutrition 3, DOI 10.1007/978-1-4614-3256-2, © Maria Lidia Herrera 2012